A
■   ■   ■
BOOK

The Philip E. Lilienthal imprint
honors special books
in commemoration of a man whose work
at University of California Press from 1954 to 1979
was marked by dedication to young authors
and to high standards in the field of Asian Studies.
Friends, family, authors, and foundations have together
endowed the Lilienthal Fund, which enables UC Press
to publish under this imprint selected books
in a way that reflects the taste and judgment
of a great and beloved editor.

The publisher and the University of California Press Foundation gratefully acknowledge the generous support of the Philip E. Lilienthal Imprint in Asian Studies, established by a major gift from Sally Lilienthal.

Urban Ecologies on the Edge

MAKING MANILA'S RESOURCE FRONTIER

Kristian Karlo Saguin

UNIVERSITY OF CALIFORNIA PRESS

University of California Press
Oakland, California

© 2022 by Kristian Karlo Saguin
Library of Congress Cataloging-in-Publication Data

Names: Saguin, Kristian Karlo, 1982– author.
Title: Urban ecologies on the edge : making Manila's resource frontier / Kristian Karlo Saguin.
Description: Oakland, California : University of California Press, [2022] | Includes bibliographical references and index.
Identifiers: LCCN 2021057713 (print) | LCCN 2021057714 (ebook) | ISBN 9780520382640 (cloth) | ISBN 9780520382664 (paperback) | ISBN 9780520382671 (epub)
Subjects: LCSH: Political ecology—Philippines—Manila—20th century. | Political ecology—Philippines—Manila—21st century. | Human ecology—Philippines. | Laguna de Bay (Philippines) | BISAC: NATURE / Ecology | NATURE / Natural Resources
Classification: LCC JA75.8 .S238 2022 (print) | LCC JA75.8 (ebook) | DDC 304.2/0917320959916—dc23/eng/20220206
LC record available at https://lccn.loc.gov/2021057713
LC ebook record available at https://lccn.loc.gov/2021057714

31 30 29 28 27 26 25 24 23 22
10 9 8 7 6 5 4 3 2 1

CONTENTS

List of Illustrations   vii
Acknowledgments   ix

Introduction: Frontiers of Urbanization   1

PART ONE
MAKING AND REMAKING A FRONTIER

1 · Birth of a Convenient Frontier   31

2 · Enclosing a Commodity Frontier   53

3 · An Unruly Frontier   80

PART TWO
THE WORK OF URBAN METABOLIC FLOWS

4 · Chains of Urban Provisioning   105

5 · Biographies of Fish for the City   124

6 · Infrastructures of Risk   140

Epilogue: Mutable Frontiers, Metabolic Futures   158

Notes   163
References   171
Index   201

ILLUSTRATIONS

FIGURES

1. Laguna Lake fisheries production, 1980–2018   *8*
2. A fishpen enclosure in the middle of Laguna Lake, 2012   *38*
3. Laguna Lake capture fisheries catch composition, 1968 and 2008   *42*
4. Laguna Lake fishpen area, 1980–1990   *58*
5. "Visiting" a fishcage aquaculture nursery (*semilyahan*), 2012   *75*
6. Main house inside a fishpen, with a view of the city in the background, 2012   *92*
7. Knifefish next to a bighead carp   *98*
8. Capture fisheries (municipal and commercial) and aquaculture production in the Philippines, 1980–2019   *110*
9. Value chain of Laguna Lake fish   *112*
10. Broker laborers or *batilyo* at the Navotas fish market, 2012   *118*
11. Transporting bighead carp, 2012   *127*
12. Fishpen production by species in Laguna Lake, 1996–2016   *129*
13. Animal protein intake in the Philippines, 1978–2008   *130*
14. Animal protein intake by income class in Metro Manila, 2008–2009   *131*

MAPS

1. Laguna Lake or Laguna de Bay and administrative jurisdiction of Laguna Lake Development Authority   *7*
2. Waterways and flood infrastructure in Laguna Lake Basin   *46*
3. Revised ZOMAP, 1999   *68*
4. Location of Navotas Fish Port Complex, major wet markets in Metro Manila, and Laguna Lake fishery centers of Binangonan and Cardona   *109*

TABLES

1. Common Capture Fisheries Gear in Laguna Lake   *39*
2. Comparison of Fishpen and Fishcage Production   *60*
3. Laguna Lake Fish Production and Fish Landings at Navotas Fish Port Complex, 2011   *107*

ACKNOWLEDGMENTS

I wish to express my deepest gratitude to a host of people who made this book's decade-long journey possible. Many thanks to Stacy Eisenstark for giving this project a chance, and to Naja Collins and the editorial team at the University of California Press for overseeing its production. In Laguna Lake, I am grateful to Marcial and Baby Valdez, Carlos Paralejas, and their families for their generosity during my stay, and to the people of Navotas and Kalinawan for unstintingly sharing their time and stories with me. I also thank Aildrene Tan, C. J. Chanco, Jose Javier, Mark Cagampan, P. J. Capio, Trixie Delmendo, Zee Alegre, and Gil Prim for invaluable assistance at various stages of the research, and the several offices, organizations, and communities in Manila that entertained my requests during fieldwork.

In Texas, I am particularly indebted to Christian Brannstrom for his unwavering support of this project from its very early beginnings as dissertation research in 2010. I thank Fiona Wilmot, Kathleen O'Reilly, Wendy Jepson, Norbert Dannhaeuser, colleagues at the Human-Environment Research Group at the Texas A&M University Department of Geography, and my College Station community. Research writing was funded by the Texas A&M University Dissertation Writing Fellowship, and a Fulbright Foreign Student Grant allowed me to study in the United States. In the United Kingdom, Colin McFarlane's guidance helped shape the initial stages of the book-writing project in 2019, along with those I met at Durham University's Department of Geography. This stage of research writing was supported by an Urban Studies Foundation International Fellowship.

I also thank the continued support and assistance from colleagues and friends based in Diliman and elsewhere, including Tin Alvarez, Yany Lopez, Andre Ortega, Trina Listanco, Vanessa Banta, and all the faculty and staff of the Department of Geography at the University of the Philippines. Subsequent fieldwork in Laguna Lake and other parts of Metro Manila was supported by university funding from the Office of the Chancellor and the Center for Integrative and Development Studies.

Early versions of certain empirical portions in the following chapters have been previously published in journals and have been expanded and updated for the book, with publishers' permission:

> Chapters 1 and 3: Saguin, K. K. (2016). States of hazard: Aquaculture and narratives of typhoons and floods in Laguna de Bay. Reprinted from *Philippine Studies: Historical and Ethnographic Viewpoints*, *64*(3–4), 527–554, by permission of the Ateneo de Manila University
>
> Chapters 2 and 3: Saguin, K. (2016). Blue revolution in a commodity frontier: Ecologies of aquaculture and agrarian change in Laguna Lake, Philippines. *Journal of Agrarian Change*, *16*(4), 571–593
>
> Chapter 4: Saguin, K. (2018). Mapping access to urban value chains of aquaculture in Laguna Lake, Philippines. *Aquaculture*, *493*, 424–435
>
> Chapter 5: Saguin, K. (2014). Biographies of fish for the city: Urban metabolism of Laguna Lake aquaculture. *Geoforum*, *54*, 28–38
>
> Chapter 6: Saguin, K. (2017). Producing an urban hazardscape beyond the city. *Environment and Planning A*, *49*(9), 1968–1985.

Finally, I thank and dedicate this work to my family—Arsenio, Helma, Kidjie, Kristine, and Yannis—and to Jake Soriano for his editorial labor and enduring patience through the years.

INTRODUCTION

Frontiers of Urbanization

ON THE MORNING OF SEPTEMBER 26, 2009, thousands in Metro Manila woke up to a sudden surge of floodwaters after hours of nonstop rain drenched the city. The tropical disturbance responsible for this record rainfall, Tropical Storm Ondoy (Ketsana), was a minor storm. It lingered several kilometers north of the megacity but drew southwest monsoon rains that dumped a month's worth of rain over a six-hour period. Manila's already overburdened urban streams and waterways failed to contain the excessive stormwater from the hills upstream, which burst their banks and inundated homes with water and mud. The city's inhabitants have long been accustomed to localized urban flooding, but the scale and damage of the Ondoy floods was unprecedented and radically altered subsequent state responses to hazards.

While images and accounts of catastrophe in the city circulated and then dissipated over the next few days—residents stranded on rooftops, motorists trapped inside vehicles, living rooms submerged in muddy water, speedboats cruising on flooded subdivision streets—those who lived along Laguna Lake's shoreline to the city's southeast had to endure flooding for several more weeks. Water that the city could not accommodate had been diverted to the lake, which rose to levels not seen in four decades. The lake's forgotten role in Metro Manila's flood control scheme as a storage space for excess stormwater quickly seeped into the public imagination again. Explanations for both the disaster and the solutions to avert future flash flooding in the city required considering the central place of the lake in making and maintaining the urban flood control infrastructure.

Four years later, in 2013, a lakeside town southeast of Manila celebrated its annual fiesta by hosting an unusual culinary contest. Competing chefs were tasked to create innovative recipes for knifefish, an exotic fish that had accidentally found and ate its way into Laguna Lake from the aquariums of urban hobbyists. The carnivorous predator posed a serious threat to commercial aquaculture in the lake, an industry introduced four decades earlier to improve fish production and meet urban and regional demands for a cheap, accessible protein source. Aquaculture enclosures eventually took hold in the lake's landscape—a contentious, transformative, and occasionally violent process—and established a lake economy that regularly supplied fish to the urban market. However, the highly invasive and voracious knifefish became a costly pest for many aquaculture producers, wiping out stocked milkfish inside the enclosures and undermining the lake's ability to provide a productive fishery.

The culinary contest was one of several attempts by the government to contain the knifefish invasion and reduce its population by demonstrating its edibility to a skeptical public wary of consuming a strange, unfamiliar fish. The winning dish, knifefish à la cordon bleu, showed that transcending the undesirability of the bland flesh and elevating the edibility of the fish body required practical and imaginative work. Fishers caught the invasive fish as a suboptimal substitute, making do with what was available in a lake ecologically transformed by the boom and bust cycles of aquaculture commodification. But due to lack of demand and limited consumption at the lake, the fish had to be brought to Manila, where its white flesh found use as an ingredient for the processing of urban street food. The exotic knifefish presented an unintended antithesis to farmed fish deliberately introduced to improve the livelihoods of lake dwellers and supply fish for the city. That both types of fish—one considered an invasive pest and the other a valuable commodity—ended up consumed as food forms in Manila shows the close and changing, intended and unexpected socioecological relations between the city and the lake in urban provisioning. It appears difficult to understand one place without the other and the resource flows that connect them.

I draw on these extraordinary and mundane scenes of conveying and provisioning to introduce the book as an urban socioecological story beyond the city. The problem of floods and food exposed urban connections that have been slowly built and maintained over time as cities expand their edges and enroll resources from elsewhere. In this book, I show how environmental trajectories of cities are inextricably tied to their frontiers, a process that simultaneously reconstitutes urban and rural spaces, ecologies, and lives. Manila embodies many of the shifting environmental challenges of the urbanizing Global South. But its proximity to the large, nutrient-rich Laguna

Lake has created particular paradoxes and conjunctures that trouble straightforward chronicles of urban development and environmental management.

Stitching together diverse accounts of the situated urban transformation of Laguna Lake in relation to Manila, *Urban Ecologies on the Edge* traces the intertwined socioecologies of the city and its urban resource frontier. In what follows, I examine the question of urban provisioning and sustenance and what kinds of work are necessary to make and maintain these relations. I engage with diverse approaches in urban, environmental, and agrarian studies to cast light on multiple accounts of urbanization as a frontier-making process that brings together natures, landscapes, and peoples across space in finding geographic solutions to urban resource challenges. By turning to the ecologies on the edge, I aim to give attention to overlooked, beyond-the-city spaces like Laguna Lake, continually made to work to produce vital resource flows that sustain city life.

Over several chapters, I weave together diverse narratives of work from frontiers to city and back: modern state plans and imaginaries of taming frontier landscapes, crisis and regulation of capitalist enclosures amid transformed lake livelihoods, lively materialities of resource frontier natures that frustrate the best-laid modern plans, access and exclusions surrounding urban commodity flows, practices of sociomaterial transformation of contradictory urban flows, and contested production of risk through flows and infrastructure. These stories have multiple trajectories that rehearse but also refuse predetermined paths of ecological transitions and take situated specificities rooted in place.

The book investigates urbanization as a frontier-making process through the case of Manila and Laguna Lake in the Philippines. Combining empirical accounts drawn from multisite fieldwork and a reading of historical materials, it seeks to provide a picture of urban socioecological transformation by engaging macroscale processes of resource flows and provisioning with the constitutive microscale practices of making a living. Through an in-depth exploration of resource frontier making in Manila, I offer a distinct political ecological approach to urbanization by drawing from a rich body of theoretical work on cities, nature, and livelihoods to describe and explain the empirical accounts across multiple sites within cities and beyond their edges. These accounts in turn are generative in helping redefine, rethink, and revise theoretical formulations of the spaces and ecologies of urbanization.

In particular, the book's framing of urbanization engages with two key concepts: frontier urbanism and urban metabolism. Both suggest that urbanization requires practical and imaginative work, whether through frontier making as the creative/destructive becoming of spaces made legible for extraction or through the delivery and maintenance of various resource flows to meet the metabolic requirements of cities. As I demonstrate through the historical and contemporary case of Manila and Laguna Lake, urban frontiers may be conceptualized as coproduced in relation to cities, molded by particular conjunctures of state power, capitalist imperatives, and everyday livelihood making. Accounting for the multiple sites of the urban by following resource flows in this case also enables rethinking urban metabolism as fundamentally driven by the work of a constellation of actors, practices, desires, and materialities that continually reshapes such relations.

Manila, with its extended metropolitan population of more than twenty-five million, became plagued with urban environmental problems throughout its rapid growth in the second half of the twentieth century. Two of its most persistent challenges—feeding its burgeoning appetite for food and water and keeping it safe from the threats of recurrent flooding—underscore its intensified dependence on resource flows from beyond its boundaries. Laguna Lake, partly due to its close proximity as a resource frontier, became an important node in state development project designs. It was imagined as a convenient frontier, a ready and pliable source of fish and domestic water and as a sink for wastes and floodwaters. As this frontier developed and resource extraction was legitimized, techniques of simplifying, erasing, and undercounting complex lake socioecologies intersected with lake dwellers' practices of dealing with ecologies and livelihoods transformed by increasing urban connections.

I focus on the political ecologies of two resource flows with particular resonance for Manila's fluid frontier urbanism and urban metabolism: fish and floodwaters. The state introduced aquaculture to spur development in the lake region while supplying steady flows of cheap fish for a growing city framed in the context of crisis in capture fisheries. It revolutionized fisheries in the lake while also changing mechanisms of property rights and initiating decades-long, conflict-ridden agrarian change rooted in deepening capitalist relations. Provisioning fish flows to the city continues to encounter multiple contradictions in both lake production and city consumption. By producing more fish for the city, aquaculture's expansion marginalized fisherfolk, the intended beneficiaries of this development project, and exposed city consumers to cheaper and more abundant but less desirable and more unsafe fish.

During the same period, the state also sought to harness the lake's water for urban domestic consumption and to manage stormwater flows in the linked Metro Manila-Laguna Lake hydrological basin. The constructed flood control network enabled the large-scale control of hydrological flows to prevent flash flooding in Manila's urban core but channeled flood flows and magnified risk for lake dwellers and their fish production. Both fish and water flows further intersect with increased waste loads that have contributed to resource conflicts that the state's various governance mechanisms had long attempted to resolve.

By following both fish and floodwaters, the book seeks to make visible the assemblages of flows, landscapes, and infrastructures—the conditions of possibility—that sustain life in the city. These configurations are simultaneously material, biophysical, and quantifiable but are also lived, imagined, and produced through work and practical activity in the everyday acts of making a living. Capital is a world-making driver of urban resource frontier making, joining with state visions and techniques to reconfigure space and nature through deepening commodification and appropriation. Yet it confronts the dynamic urban edge in emergent ways, producing a politicized zone where lives and landscapes fight back, realign, or refuse their frontier making. Through these fluid stories set in Manila and Laguna Lake, the book extends an understanding of how urbanization produces particular, often paradoxical, ecologies in cities, edges, and beyond, and who wins and loses in the process of urban environmental change.

FLUID URBANISMS: MANILA'S FISHBOWL AND TOILET

Manila, often used to refer to the broader Metropolitan or Metro Manila urban region, sits on a narrow stretch of coastal, alluvial, and hilly volcanic land with water on two sides.[1] To the west lies Manila Bay and its deep harbor, which has played a vital role in Manila's history as one of the first global cities. Manila was a colonial port city that connected Asia and Europe, a central node in the Spanish Empire's territorial and economic expansion from the sixteenth to the nineteenth centuries. Located near the point where the Pasig River meets Manila Bay, the City of Manila is the highly dense, old core of the metropolis, expanding from a precolonial coastal urban settlement to a colonial capital socioracially divided by a fortification.[2]

To Metro Manila's southeast lies Laguna Lake or Laguna de Bay, a shallow freshwater lake whose significant role in Manila's city making is much less recognized and whose urban connections are less visibly obvious (see map 1).[3] Upon gaining independence from formal American colonial rule (1898–1946), the Philippine state embarked on various development projects that were increasingly oriented to the urban needs of an expanding Manila. Laguna Lake served as a proximate source for many vital resource needs, including food, water, and drainage and wastewater management, initiated primarily by the state body Laguna Lake Development Authority (LLDA) (see map 1).

At around 90,000 ha (900 km^2), the lake is the largest in the Philippines and the third largest in Southeast Asia. Twenty-one rivers in its watershed drain into the lake, but the Pasig River, which cuts across urban Metro Manila, is its only outlet to the sea. The river brings saline backflow, alongside urban pollution, to the lake from Manila Bay during drier seasons when the lake's water levels fall below sea level. As a result of the prehistoric collapse of a volcanic caldera, the lake's 250 km shoreline follows a hoofprint-like configuration, with two peninsulas dividing the lake into three lobes (East, Central, and West Bays) that have temporally differing levels of salinity. The lake is cut in half by Talim Island, a long, jagged, volcanic land mass separated from the mainland by the Diablo Pass, which at 20 meters is the deepest section of the lake.

The lake is highly eutrophic due to the abundance of nutrients that encourage the growth of phytoplankton. During the transitional period between the dry and wet seasons in May-June, algae blooms temporarily turn the dull water a deep shade of emerald green. This hypereutrophic property served as one of the primary justifications for the state's introduction of extensive aquaculture, enabling the growth of fish even with very minimal external inputs. The lake's shallow depth at 2.5 meters also facilitated construction of aquaculture enclosures, as fences can easily be staked to the muddy bottom. These limnological processes have historically supported capture fisheries in the lake, and since 1970, aquaculture production. As the blue counterpart to the green revolution, aquaculture embodied the parallel aims of improving food production through technological and institutional changes. Laguna Lake pioneered extensive, commercial aquaculture based on a body of water, and its contribution to urban fish diets has become so significant that the lake has been termed Manila's "freshwater fish bowl" (Lasco & Espaldon, 2005, p. 39).

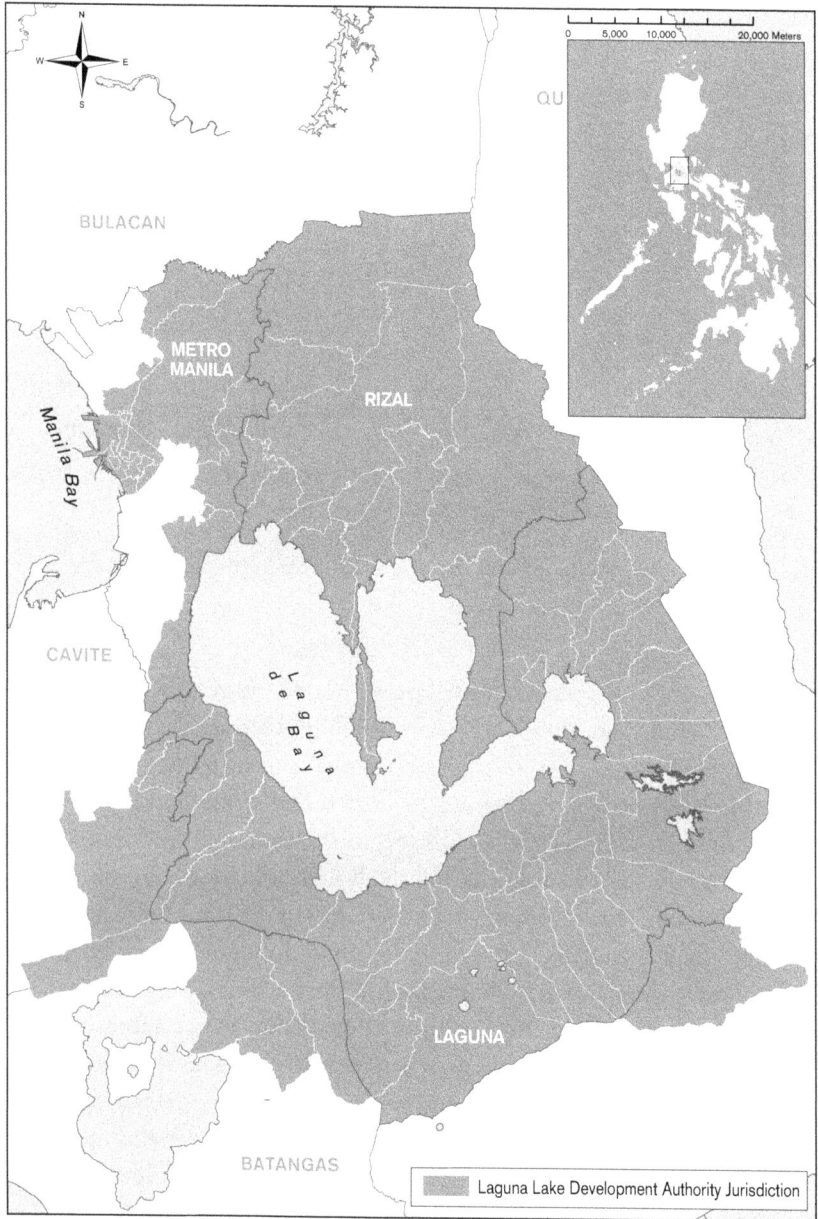

MAP 1. Laguna Lake or Laguna de Bay and administrative jurisdiction of Laguna Lake Development Authority. Map by Patricia Anne Delmendo.

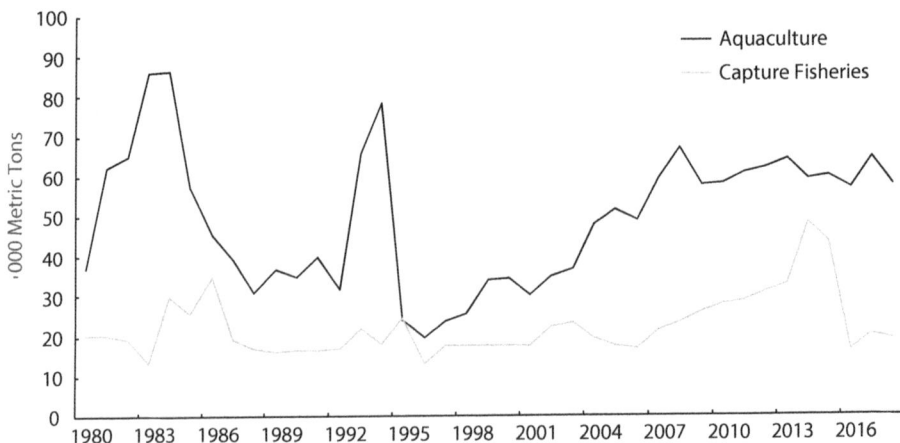

FIGURE 1. Laguna Lake fisheries production, 1980–2018. *Sources:* Laguna Lake Development Authority (1995b); National Statistical Coordinating Board (1999); Philippine Statistics Authority OpenSTAT database. *Note:* Information on capture fisheries production between 1997 and 2001 is unavailable from the database and is presented as the average of preceding and succeeding years.

Aquaculture production in the lake surpassed capture fisheries' production only a few years after it was introduced, peaking at 85,000 metric tons in 1985 (see figure 1). Among the low- to mid-value introduced fish species, milkfish (*Chanos chanos*), tilapia (*Oreochromis niloticus*), and bighead carp (*Hypophthalmichthys nobilis*) are the three most commonly produced.[4] They are grown in large-scale fishpens and small-scale fishcages, aquaculture production systems that together occupy a seventh of the lake's total area.

More than five million people reside along the shores of the lake, with at least three thousand directly engaged in small-scale cage aquaculture and thirty-five thousand fisherfolk still making a living from capture fisheries using various active and passive gear (Israel, 2007).[5] The resulting livelihood mosaic in the lakeside villages is complex, in which traditional capture fisheries production, aquaculture production, and other activities continue to be shaped by urbanizing processes in Metro Manila and surrounding regions.

The Metro Manila and Calabarzon regions form the country's urban and industrial core, accounting for half of the total gross domestic product and two-thirds of manufacturing employment and output (Shatkin, 2008). Metro Manila's urban landscape and built environment have expanded both vertically and horizontally, driven by a variety of processes including in-migration, neoliberal restructuring in governance, and transnational flows, sprawling over rural, transitional, and mixed land uses (Garrido, 2019; Kelly, 2000; Kleibert & Kippers, 2016; Ortega, 2016; Shatkin, 2005, 2008). The nearby Calabarzon region, which surrounds much of Laguna Lake, has been similarly urbanizing, facilitated by the Calabarzon Project, a regional industrial development plan covering the provinces of Cavite, Laguna, Batangas, Rizal, and Quezon. The project has led to dramatic transformation of areas around Laguna Lake and its watershed, such as agricultural land conversion, displacement, and in-migration, as well as increased pollution emissions, the ecological impacts of which are felt in the lake as a sink for the wastes produced by these activities (Canlas, 1991; Kelly, 2000; Lasco & Espaldon, 2005; Ortega, 2012).

Despite decentralization attempts, Metro Manila's population continues to grow, from two hundred thousand at the turn of the twentieth century to more than ten million by the turn of the twenty-first century. Yet this growth in numbers conceals wide inequality and spatial fragmentation in the city that harkens back to the colonial division represented by the earlier urban wall. Nearly three-quarters of the urban population belongs to the lower and extremely lower socioeconomic classes, with the proportion of urban population residing in poorly served slums ranging from a fifth to half throughout the latter half of the twentieth century (Arcilla, 2018; Arn, 1995; Ortega, 2016; Shatkin, 2005). Manila is highly fragmented, and these inequalities have expanded spatially to the city's edge and temporally toward an uncertain environmental future as a disaster-prone metropolis where earthquakes, typhoons, floods, and pollution hazards pose recurring threats that affect city dwellers unevenly. Its expansion has constantly put a strain on its ability to meet its resource needs, historically addressed by the state by constructing networks of provisioning and sustenance that stretch beyond the borders of the urban region.

URBANIZATION ON THE EDGE

This book traces the resource flows that sustain Manila through its relations with Laguna Lake, its convenient frontier, and lays bare the multiple political ecologies that constitute these flows and the frontier. I turn to the polysemy of the phrase "on the edge" in its multiple meanings to situate these relations. "Urban ecologies on the edge" invokes at least three senses: a location, a relation, and a condition.

As a location, the edge refers to the urban fringe, the zone where the city dissolves into the beyond-the-city. This urban periphery is more of a continuum or gradient than a geographical area with an abrupt or static boundary. It is more gradual, patchwork, hybrid, or ambiguous than delineated, often shaped by a mix of multiple urban and rural processes and logics, characterized by situated transformations and unpredictable juxtapositions. The political ecologies of these dynamic and transitional spaces on the urban edge have been framed in distinct but related, sometimes overlapping, and contradictory terms: the peri-urban (Bartels et al., 2020; Myers, 2008; Simon, 2008), suburban (Keil & Macdonald, 2016; Ortega, 2012; Pares et al., 2013), exurban (McKinnon et al., 2019; Walker & Fortmann, 2003), and megapolitan (Gustafson et al., 2014). Yet edges take diverse historical-geographical forms, extend beyond the usual hinterland borders, and are situated in differing contexts, necessitating attention to their dynamic interplay. Laguna Lake is a particular example, as it sits on Manila's expanding edge, the built environment of the city literally stopping at the lakeshore, even as its urban connections, flows, and impacts extend far beyond.

As a relation, the edge denotes limits, transitions, and liminality, being wedged between two worlds: in between the core and its margins, the city and its frontiers. The edge reflects a spatiotemporal relation manifested in particular times and places, suggesting that the history and fate of places like Laguna Lake and Manila are imbricated relationally through urban processes. The in-betweenness creates unique and novel ecological relations that require a focus on both city and frontier and their liminal edges. Ecologies describe the multiple relations between individuals and their physical environment, relations that are more accurately referred to as socioecological. On the edge, the socioecological is more visibly constitutive of the production of both city and frontier.

Finally, being on the edge alludes to a condition of uncertainty, precarity, and being unsettled. Talking about ecologies on the edge suggests socioecological relations and transformations are marked by dynamic shifts and surprises, with the looming sense of being on the precipice of transforming into a different state. There is a degree of undecidability and provisionality in the kinds of arrangements emerging as diverse spaces, peoples, and ecologies are juxtaposed (Massey, 2005; Roy, 2016a; Simone, 2020). The term is an appropriate description of certain processes of urban frontier making in Laguna Lake and other edges, where visions of space and material transformations reconfigure precarious lives and landscapes, which in turn redefine trajectories in unexpected and never complete or predetermined ways.

Urban edges are relational and may occur in various formations, from the proximate peri-urban frontier such as Laguna Lake to further resource hinterlands connected by extending capital flows and globalized infrastructure networks. They are characterized by a type of "edginess," whose diverse emergent politics and ecologies require further exploration as they are situated in place. In this book, I explore one of these formations rooted in a particular place at a particular moment, but I aim to keep the relational tension between city and frontier in focus to think about other cities and frontiers elsewhere. Framing the ecologies of edginess engages with two concepts with rich histories: urban metabolism and the frontier.

URBAN METABOLISM AND THE POLITICS OF FLOWS

Urban metabolism—a boundary concept mobilized in multiple ways in both the natural and social sciences—anchors the socioecological exploration of urbanization on and beyond the edge. In its organicist sense, as understood in industrial ecology, it presents an idea of the city as supplied by flows of materials and energy from the outside necessary for the city's continued functioning. Employing a systems approach, scholars in this field argue that quantifying and measuring resource flows, stocks, inputs, and outputs is a necessary precondition for planning toward urban sustainability (Kennedy et al., 2007; Pincetl et al., 2012).[6]

But as critical urban scholars have pointed out, flows and their infrastructure also bear deep-rooted histories, situated practices, and contested politics that require casting attention to constituted social relations and lived experiences. Scholarship in the field of urban political ecology (UPE) has deployed a historical and political understanding of urban metabolism drawing from Marx's original use of the term.[7] *Metabolism* becomes a metaphor for the material and symbolic production of nature in cities through circulation, exchange, and transformation, as well as the co-constitution of social labor and material processes in capitalist urbanization (Heynen et al., 2006; Heynen, 2014). Urban political ecologists emphasize socionatural relations through a historical and political approach to the production of urban natures, wherein both cities and nature are understood as coproduced or as hybrids that bring together heterogeneous actors and objects (Gandy, 2004; Swyngedouw, 2006).

The urban metabolic and socionatural transformations of city and beyond-the-city spaces are inherently political questions. Urban political ecological work is thus explicitly concerned with transforming unjust urban relations by revealing what is hidden or made invisible in the capitalist urbanization of nature (Heynen et al., 2006). It brings empirical attention to control and access to metabolic flows, which benefit a group of people or particular places at the expense of others, showing how urban socionatures are constituted by social power as a result of attempts by various groups to mobilize their interests and access resources (Swyngedouw, 2004).

This book focuses on the material and imaginative politics inseparable from the production of socionatural transformations, tracking material flows as in industrial ecology to show how the ecological connections between the city and its frontier matter (Demaria & Schindler, 2016; Newell & Cousins, 2015). It also explores the materiality of nature in urban metabolism in its multiplicity and grounds metabolism by illustrating the various ways that ecologies are urbanized through practical acts of work and labor. It seeks to maintain the tension between a microscale focus on individual and collective practices and ways of seeing and a macroscale transformation and control of flows.

Flow is an important metaphor to describe metabolism's spatial dynamics. It implies fluid movement and circulation, which are not simply material but are constituted through various relations in the process of flows maintenance (Kaika, 2004; Swyngedouw, 2006). Urban metabolic transformations involve circulation of commodities as "forms of metabolized hybrid socionatures" (Swyngedouw, 2006, p. 109) that are produced under exchange value relations. Water has been the subject of many UPE urban metabolism narratives in various cities (and similar framings have been applied to other urban flows such as alcohol, fat, and wastes).[8] Yet apart from Susanne Freidberg's (2001a) notable work in Burkina Faso, the circulation of food supplied from beyond the city has largely been overlooked by urban political ecologists. Unlike water channeled to the city, food produced elsewhere often requires particular material and symbolic transformations of landscapes and flows and brings together a host of actors, places, and relations before being consumed.[9]

Because the city sources most of its food from outside, city dwellers consume food primarily as commodities via market exchange mediated by increasingly global supply or value chains. Food is metabolized through various practices and work at different sites as it moves toward and around the city, as commodification fundamentally transforms people's relationship with nature. Exploring the displacement and geographical lives of food commodities through metaphors such as chains or networks presents a point of productive engagement with urban metabolism, examining how commodity flows constitute urban natures through everyday practices of provisioning and securing livelihoods (Castree, 2004; Cook et al., 2006; Hughes & Reimer, 2004).[10]

Other types of flows, on the other hand, are managed because they present harm and risk to urban populations. In the modern Western world, "bad" water, considered harmful and hazardous, is expelled from private domestic spaces and hidden in public city spaces (Kaika, 2004; Karvonen, 2011; Walker et al., 2011). Wastewater and stormwater flows are the noncommodified and unwanted opposite of municipal water (good water) or of the vital inflow of food and energy. As an undesirable hazard, bad water in cities is often rationalized, displaced, and efficiently conveyed elsewhere through modern infrastructure networks (Karvonen, 2011). The sanitary city and the networked city emerged as twentieth-century visions of the modern city that sought to expel and control metabolic flows through integrated infrastructural services and initiating changes in the built environment (Gandy, 2004; Graham & Marvin, 2001; Melosi, 2008). Yet in many Global South cities, these flows often frustrate or overcome technocratic managerial attempts at control through engineering solutions, resulting in spatial fragmentation manifested in uneven exposure to destructive hazards (Collins, 2009; Mustafa, 2005; Schramm, 2016).

A metabolic lens applied to multiple resource flows that sustain and constitute urban life suggests that urbanization assembles diverse things, relations, and politics in making and maintaining particular socioecological arrangements across space. Cities are places always in the making (Lepawsky et al., 2015; Simone, 2010), and city-making processes are also located in the everyday material and symbolic practices surrounding resource flows and transformations. This situated everyday urbanism (Lawhon et al., 2014) stretches across space from cities to their frontier, as material transformations of flows intersect with inhabitants' understanding and experience of urbanization, including their acts of doing and making a living situated in place. An emphasis on metabolism beyond the city is also politically generative, as it extends "the potential sites of interventions" and widens "the objects of analysis and the epistemology of social change" (McFarlane, 2013, p. 500) within both visibly politicized landscapes and hidden ecologies embedded in the broader geographies of power (Huber, 2017).

RESOURCE FRONTIER URBANISMS

Connections that make cities and the spaces they transform have been framed as different binaries—urban-rural, city-countryside, city-hinterland, core-periphery, agglomeration-operational landscapes—and emphasize the multiple sites where the urban's constitutive outside resides (Reddy, 2018; Roy, 2016b). These relations capture the simultaneous marginality and centrality of these spaces in urbanization: spaces that are peripheral yet vital to city making. To understand city-making processes and the geographies of urban metabolism, we need to grasp how these spaces and ecologies also contribute to making the urban. In dominant models of urban and economic geography, and in industrial and ecological economic analyses, the city is a distinct spatial entity from its hinterland or frontier (Gandy, 2004; Golubiewski, 2012; Mehzabeen, 2019). Relational approaches to cities, in contrast, map the multiple spatialities of the city and urbanization in and beyond the bounded agglomerations in which they are often represented (Lepawsky et al., 2015; Massey, 1994; Reddy, 2018).

The frontier, as I show in this book, opens up possibilities for relationally understanding urban spaces transformed beyond the city. However, the term carries conceptual baggage as an overdetermined category, requiring specificity in its usage and sensitivity to its situated and diverse forms. Frontiers are historically and geographically specific, representing a moment that invokes particular assumptions about center and margins rather than a self-evident concept that manifests uniformly or timelessly. The largely rural-oriented body of work on resource frontiers serves as a starting point to think through the kinds of spatial and ecological forms and processes that urbanization produces as it moves resources between cities and frontiers.

The frontier denotes a dynamic spatiality. Classic works suggest linear movement or a mobile front as frontiers expand into marginal spaces over time, such as Frederick Jackson Turner's (1920) frontier thesis on the history of the American West and the march of civilization. Frontiers are also often understood as political zones designed in relation to the state, where sovereignty is spatialized and encounters the state's territorial limit (Saraf, 2020; Watts, 2018). Yet frontiers are more than just a timeless spatial category or place whose boundaries can be demarcated on a map. Rather, they are emergent historical products, produced in relation to the center, as something that *takes place* (Rasmussen & Lund, 2018). While frontiers have historically been understood in terms of particular spatial imaginations such as empty rural lands at the remote margins, when reframed relationally, they may exist in areas such as Laguna Lake, proximate to the core and populated centers (Barney, 2009; Fold & Hirsch, 2009; McGregor & Chatiza, 2019; Pullan, 2011; Rasmussen & Lund 2018).

Frontiers are mutable and mobile (Cronon, 1996), but their continuous formation is cyclical rather than linear as they emerge and vanish, move and return (Cons & Eilenberg, 2019; Rasmussen & Lund, 2018). *Frontier making* captures this relational dynamism, emphasizing the becoming of frontier as an ongoing process. *Resource* frontiers meanwhile signal the creative/destructive incorporation of margins into the orbit of state and capital, emerging at particular moments when a new resource becomes amenable to extraction and commodification (Cons & Eilenberg, 2019; Rasmussen & Lund, 2018; Tsing, 2005). As sites where the state's territorial power and modern visions of order and capital's logic of accumulation intersect with lives and landscapes, resource frontiers are dynamic spaces of conflict, change, and potentialities. Resources and frontiers are co-constituted in the process of the discovery and release of natural resources, reconfiguring existing livelihood-ecological relations.

In her expansive work on resource frontiers, anthropologist Anna Tsing (2005) identified a few key features of frontiers: imaginative, liminal, unmapped, and lively. First, as an imaginative project, frontiers are discursive constructions produced at specific moments in time, suggesting a particular relationship between core and margins (Cons & Eilenberg, 2019). Frontiers are shaped by contradictory visions of what is and what might be. In this imagination, places are framed as empty, wild, untouched, and underdeveloped but at the same time full of potential and opportunities, primed for exploitation, resource extraction, and state intervention (Bridge, 2001; Cons & Eilenberg, 2019; Eilenberg, 2014). Modern ideologies of the nation, development, and progress clash with other knowledge practices, materially shaping both places and processes (Fold & Hirsch, 2009; Rasmussen & Lund, 2018).

Second, as a contact zone and "an edge of space and time: a zone of not yet" (Tsing, 2003, p. 5100), frontiers are best understood as transitional and liminal. Often conflated with borderlands, suggesting a similar interstitiality in the margins of state power, frontiers are a zone of interface where two worlds meet and confront each other, as spheres of friction where negotiation and collaboration play out (Tsing, 2005). Frontier liminality is both spatial and temporal, as in-between places, emerging amid or beyond state space, in the process of transition into something else in the future (Fold & Hirsch, 2009; Rasmussen & Lund, 2018). But frontiers are indeterminate and unpredictable, as they emerge through chance and contingency, sometimes veering off from plans and intentions and often without coherence or consistency (Cons & Eilenberg, 2019; Tsing, 2005).

Third, as a form of unmapping, frontier making requires erasure of existing relations to establish new configurations of rule. Through representational and practical techniques, frontier making simplifies the situated complexities of places and abstracts local knowledge and livelihoods from their past entanglements (Tsing, 2005). Unmapping requires the undoing of established order and expunging all other claims to institute a new regime for the purpose of legitimating and facilitating resource extraction and exploitation (Cons & Eilenberg, 2019; Saraf, 2020). Technologies that territorialize for state legibility are central to this process, such as maps, fences, and titles, codifying new understandings of space and nature (Li, 2014; Peluso & Lund, 2011). The unmapping and effacing of previous orders is often characterized by violent encounters as states attempt to discipline and instill control in frontiers, dispossessing and excluding local people (Cons & Eilenberg, 2019; Rasmussen & Lund, 2018). Frontier making as unmapping is necessarily accompanied by territorialization as a remapping and reordering of space (Peluso & Lund, 2011; Rasmussen & Lund, 2018).

Fourth, frontiers are lively in the sense that they are populated with lives, livelihood, and liveliness. Despite attempts to unmap, abstract, simplify, and efface, this liveliness plays an active role in the making of frontier and haunts as an unexpected surprise that subverts neat plans of control (Cons & Eilenberg, 2019; Mitchell, 2002; Tsing, 2005). The entanglements of human and nonhuman lives and materialities shape frontier trajectories and are reconfigured through new political subjectivities and resistance against frontier making (Saraf, 2020). Frontiers are populated by multiple actors, processes, and configurations and have thus been framed as assemblages and constellations to emphasize their diversity and conjunctures (Cons & Eilenberg, 2019; Eilenberg, 2014; Li, 2014).

Frontiers play a crucial role in sustaining and reproducing capitalist relations and accumulation. Indeed, capitalism is characterized by frontier movements, as capital colonizes the uncolonized in both processes of commodification (expanding capitalist relations to deliver more commodities) and appropriation (bringing the noncommodified realm closer to keep capitalist production costs down) (De Angelis, 2007; Moore, 2015). Frontier sits at the boundary between the commodified and the uncommodified, a boundary that capital seeks to transcend in search of spatial solutions to its internal systemic crises

(Harvey, 2003, 2006; Moore, 2011). Primitive accumulation and accumulation by dispossession, key moments in the development of capitalism, find their spatial manifestations in frontiers, characterized by enclosures that cleave people from their means of production and by multiple processes that destroy and expropriate the commons (De Angelis, 2007; Glassman, 2006; Harvey, 2003; Rasmussen & Lund, 2018). It is in this context that environmental historian Jason Moore (2015) reframes the frontier as a *commodity* frontier vital to capitalism's reproduction not only in expanding spaces of capitalization but in extending zones of appropriation to deliver cheap food, labor, energy, and raw materials to overcome limits of commodification as barriers to capitalism's expansion.

Urbanization is often overlooked in geographical accounts of frontier making, which have largely been interested in examining shifting forms of contestation and extraction in remote, rural lands. A few commentaries, however, have alluded to peri-urban spaces as resembling frontier zones of transition and to how urban processes have blurred frontiers and centers, questioning distance and remoteness as defining features of frontiers (Cons & Eilenberg, 2019; Fold & Hirsch, 2009; Gururani, 2020; McGregor & Chatiza, 2019; Pullan, 2011; Rasmussen & Lund, 2018). But as a flexible concept that describes the relationship between the center and the margins, frontiers and urbanization can be mapped onto each other as relations between cities and hinterlands are co-constituted by resource production and flows. Frontier making tied to (capitalist) urbanization creates stories of spaces reconfigured for urban resource needs. An urban resource frontier promises progress and development for frontiers, achieved through increased integration with the city to solve its urban resource problems, facilitated by state territorialization. As rich and diverse as existing resource frontier accounts are, they often stop short of tracing resource flows beyond the frontier, of what happens to resources as they travel and circulate, and how in turn they (re)constitute the center. This is the frontier story that this book seeks to tell through urbanization, building on a parallel tradition in urban political ecology.

Urban political ecologists have worked within a relational understanding of cities as constituted by a plethora of multilayered flows, suggesting that "there is no longer an outside or limit to the city" (Swyngedouw & Heynen, 2003, p. 899). Urbanization therefore "produces both a new urban and rural socio-nature" and constant "extension of urban socioecological frontiers" (Swyngedouw, 2006, p. 114). Work on planetary urbanization attempts to challenge UPE on its unfulfilled claims of bringing these frontiers into urban narratives through a critique of its methodological cityism or its primary focus on cities as artifacts of urbanization (Angelo & Wachsmuth, 2015). However, UPE works mapping the urbanization of resources have long demonstrated the material and symbolic coproduction of cities and frontiers through uneven metabolic processes (Connolly, 2019).[11]

Erik Swyngedouw's (2004) account of the urbanization of water in Guayaquil in Ecuador, for example, combines historical and ethnographic approaches to link past colonial processes with contemporary patterns of uneven urbanization, linking agrarian landscapes with the city. Social power permeates control of where water flows, resulting in the highly uneven distribution of access to abundant potable water among urban dwellers. Matthew Gandy's (2003) urban environmental history of New York City similarly examines how resources from the city's hinterland were reworked for the city by tracing their flows, politics, and ideologies. Both accounts have been inspired by William Cronon's (1991) environmental-economic history of Chicago and its hinterlands, which narrates stories of city-frontier coproduction through commodity flows of grain, lumber, and meat. Chicago's urban history is told through its constitution by resource flows from the fields and forests of the Midwest, emphasizing their coconstitution and interdependence: "They created each other, they transformed each other's environments and economies, and they now depend on each other for survival" (Cronon, 1991, p. 384). It seems impossible therefore to discuss urban ecologies without including these metabolic flows and city-frontier relations.

These urban works suggest that capitalist relations embodied by contemporary cities configure the urbanization of nature (Swyngedouw & Heynen, 2003), wherein the city-countryside dynamic becomes a spatial relation of the logics of capital (Moore, 2011) and reflects historical moments in capitalist development that reinforce particular ideologies (Williams, 1973). The city/urban and the noncity/rural are co-constituted in urban metabolism (Harvey, 1996), not in a sense of metabolizing other places but in that various practices and relations constitute this metabolism. The capitalist production of urban space requires a corresponding production of nature, initiated through concrete practices of laboring, a transformative act that brings together the human and the nonhuman (Braun, 2005; Gandy, 2004; Loftus, 2012; Smith, 2008). Yet it needs to be recognized that, as scholars drawing from postcolonial readings argue, capitalist urbanization is a significant but not the only force that shapes the urban frontier (Jazeel, 2018; Reddy, 2018; Roy, 2016b).

Tracking urban transformations beyond the city to the multiple sites of frontier urbanism shows that landscapes, bodies, and communities are shaped by urbanization in radical and banal ways (Arboleda, 2016; Ghosh & Meer, 2021; Gustafson et al., 2014; Hommes & Boelens, 2017; Kanai, 2014; Lepawsky et al., 2015). The urban frontier reveals realms and relations that have been traditionally the subject of agrarian studies and (rural) political ecology, with themes such as access, power, and control surrounding livelihoods, enclosures, and socioecological change similarly applicable to urban formations (Bartels et al., 2020; Cornea et al., 2016; Robbins, 2011). Despite moves to rethink the epistemologies of the urban by challenging the fixity of the city (Brenner & Schmid, 2015), the city remains analytically useful in urban frontier-making explorations, as it enables us to mark historical shifts in metabolic relations tied to decisions made about and for these spaces (Connolly, 2019; Davidson & Iveson, 2015; Rickards et al., 2016). The urban frontier is urban "because of [its] relation to unfolding processes of city-making" (Davidson & Iveson, 2015, p. 655), requiring continuous explanation of the processes through which the urban is made (Roy, 2016a).

The frontier and all its conceptual heft help us think of urbanization's margins and edges in a relational way as zones of transition that take diverse geographical forms. But it also trains our attention to the specific features in these margins and edges that are shaped by urbanization, including their imaginative, liminal, unmapped, and lively characters. Despite its breadth, frontier is insufficient on its own to understand the socioecological extent of the urbanization of nature, especially as resource flows from the margins reconstitute the city. Thus, there is a need to keep these spaces in a relational tension with the city, turning attention to how resources and metabolic flows co-constitute both city and frontier.

URBAN ECOLOGIES: MATERIALITY, INFRASTRUCTURE, AND PRACTICES

The ecological history of cities may be interpreted through their need to continually transform nature and extend frontiers further (Swyngedouw, 2004). Three key concepts that constitute socioecological coproduction in these spaces are vital in frontier making and mediating metabolic flows: materiality, infrastructure, and everyday practices.

First, metabolic processes and circulation of flows encounter nonhuman natures that are neither inert nor passively acted upon by humans as the agent of transformation. Nature's materiality, or the "ontological existence of those entities we term 'natural' and the active role those entities play in making history and geography" (Castree, 1995, p. 13), is central in how we conceptualize urban ecologies within and beyond the city. The matter of nature (FitzSimmons, 1989) and its place in frontiers have been extensively explored in fields such as agrarian political economy, where the role of nature and its recalcitrant materiality in capitalist production focuses on natural obstacles to capitalism in nature-based industries (Banoub et al., 2020; Boyd et al., 2001; Goodman et al., 1987; Henderson, 1999; Kloppenburg, 2005; Mann and Dickinson, 1978). Work in this tradition argues how capital overcomes, circumvents, or takes advantage of the problem that nature poses in agriculture and similar industries, while considering implications for institutions, regulation, scale, and dispossession (Banoub et al., 2020; Bridge, 2000; Huber & Emel, 2009; Sneddon, 2007).

Talk of materiality of nature matters as it shapes social relations of production, including the organization of labor processes, institutions, and relations between producers (Benton, 1989; Mann, 1990; Prudham, 2005). Producing water or through water, for example, encounters material properties distinct from land-based production, such as fluidity, circulation, and the complex biotic/abiotic factors that comprise water quality (Bakker, 2004; Mansfield, 2004; Sneddon, 2007). Perishability and freshness have historically shaped trajectories of food production, distribution, and consumption under capitalism (Freidberg, 2009), as have animals as lively capital (Barua, 2019). Beyond political economy and capitalist natures, materiality has also been deployed through the lenses of cultural studies of commodities, corporeality, actor networks, assemblages, new materialism, and other relational ontologies (Bakker & Bridge, 2006; Bennett, 2010; Miller, 2005; Peters, 2012). Materiality through these lenses yields novel understanding of socionatures at work in resource production, posing ontological challenges to nature/society by emphasizing hybridity, performativity, multiplicity, and relationality of the material in stories of environmental change (Bakker & Bridge, 2006).

In urban political ecology, the materiality of urban nature is understood in terms of the socionatural hybridity—simultaneously social and natural—permeating urban metabolisms. The "hybrid" or "cyborg" metaphor brings together nondualist views of nature and society that claim cities do not just have an ecological dimension but are instead constituted by ongoing transformations that coproduce both urban and rural socionatures (Gandy, 2005; Swyngedouw, 2006). Combining historical materialist and new materialist approaches to the materiality of nature has inspired an understanding of the urbanization of nature as an ongoing process of bringing together heterogeneous objects with consequences for thinking seriously about the place of more-than-humans in city and frontier making and how materiality makes a difference in urban metabolism (Demaria & Schindler, 2016; Holifield, 2009; McFarlane, 2011; Ranganathan, 2015; Swyngedouw, 2006).

Second, infrastructure as a sociotechnical element of urban metabolic relation facilitates bringing resource flows to the city and connecting the city with the frontiers that sustain them (Loftus & March, 2016; McFarlane & Rutherford, 2008; Monstadt, 2009; Silver, 2015). As networked infrastructures and urban forms coevolve, extending further out with greater spatial reach, they present central nodes within which contestations surrounding access and politics take place both ideologically and materially. Infrastructure facilitates frontier making by integrating adjacent and distant places through resource flows and the complex ecologies that make these possible (Carse, 2012; Furlong & Kooy, 2017; Graham & Marvin, 2001).

Infrastructure serves as the underlying or background mechanism that enables the work of circulation of flows of things, resources, goods, people, and ideas across space (Larkin, 2013; Star, 1999). As hard, rigid structures, they appear as solid, durable, and permanent fixtures of the landscape but simultaneously require repair and retrofitting to continue functioning, meet new demands, and resist being outmoded or obsolete (Howe et al., 2016). Embodying visions of modernity and control of nature, many of these infrastructure networks become vital sites in the struggle for access to flows but whose very political character is often rendered technical and invisible (Graham & Marvin, 2001; Kaika & Swyngedouw, 2000; McFarlane & Rutherford, 2008). They appear utilitarian but are inherently political, reflecting and embodying structures of power (Graham, 2010). Their absence or fragmentation in space is as contentious as their visibility and presence.

Urban infrastructures reflect ideals of modernity and create the grounds for the operation of resource frontiers. Consequently, they also become sites and objects that are the visible target of resistance when they convey harmful or disrupted flows as socioecological burdens for people. Thinking about infrastructures requires recognizing their paradoxical character (Howe et al., 2016) but also the ways that they are lived, experienced, and seen (Graham & McFarlane, 2014; Simone, 2004), locating them beyond the Global North ideal of functioning centralized networks and within the realm of everyday practices (Furlong & Kooy, 2017; Lawhon et al., 2014; Monstadt & Schramm, 2017). Infrastructure becomes an important site for urban metabolism materially, ideologically, and politically.

Finally, the ecologies of urbanization are also situated within practices and experiences of inhabitants in both cities and beyond their edges. The urban is lived, inhabited, and experienced—within and beyond processes of capital accumulation—and situated within particular conjunctures (Roy, 2016a). The everyday becomes a key site of socioecological transformation and is a product of local contingencies that vary across contexts (Connolly, 2019; Doshi, 2017; Lawhon et al., 2014; Simone, 2019). Situating and grounding urban metabolism through accounts of the everyday in Global South cities produces an alternative or counterpoint to how we understand the urban (Lawhon et al., 2014). It involves rethinking how resources are constituted, diversifying narratives of which flows and transformations matter in particular places. The metaphor of flow that defines the fluidity of resource movement may similarly be reframed as constituted by ordinary practices of doing that produce and reproduce social power, urban difference, and space (Doshi, 2017; Lawhon et al., 2014; Zimmer, 2015).

Accounting for practices surrounding transformation of resource flows at various sites in frontier urbanism demonstrates the generative acts of gaining or restricting access, positioning, meaning making, and material change as people reshape ecologies of connection. These extend to diverse, mundane acts of doing that rework the urban environment and sustain obdurate relations in minute and mighty ways. Metabolism of food, for example, is not just a question of urban-rural exchange of nitrogen or phosphorus but is also the situated relations surrounding its production, circulation, and consumption. Materiality, infrastructure, and situated everyday practices all point to the labor necessary to build and maintain socioecologies. They present opportunities to engage various readings of the city and the frontier linked by urban metabolic relations.

FOLLOWING THE FLOWS AND ORGANIZING THE NARRATIVE

The urban ecologies on the edge encompass multiple sites between Manila and Laguna Lake and extended time frames from the past to the contemporary. I have sought to capture these stories through accounts from multisite fieldwork research complemented by a reading of documents from state and scientific reports and news articles published from 1905 to 2017. The multisite methodological strategy enabled me to follow metabolic flows of fish and the geographical lives of a commodity along various nodes (Cook et al., 2006; Freidberg, 2004; Ribot, 1998), as well as to map the flows of floodwater mediated by infrastructure networks. This strategy allowed placing a geographically extensive process involving various agents in specific sites, paying attention to macro processes that constitute the context while allowing for a flexible engagement with theory through empirical cases. Multisite approaches identify the diverse sites of the urban and the diffuse practices that constitute them (Lepawsky et al., 2015), linking together places shaped by similar processes (Freidberg, 2001b).

Through the multisite research strategy, I have covered a wide range of livelihood engagements within and beyond the fish value chains and infrastructure networks that connect Laguna Lake and Metro Manila. Anchored on a "follow-the-thing" approach focused on fish flows (Cook et al., 2006), I conducted semistructured interviews with a total of 115 fish producers, traders, consumers, and lake residents, as well as state officials and key actors in Manila and in Laguna Lake. The bulk of the interviews, participant observations, and field research in the lake took place in the first half of 2012 in two Rizal villages (*barangays*): Navotas in Cardona municipality and Kalinawan in Binangonan municipality. These villages, located within the primary fisheries municipalities of the lake, were selected because they represented differing engagements with fisheries and aquaculture. Navotas hosted a diverse set of fishing-based livelihoods, while Kalinawan is almost exclusively dependent on cage nursery aquaculture, allowing for comparison of diverging agrarian trajectories of urban-oriented aquaculture.

In Manila, I conducted interviews and observations at the Navotas Fish Port Complex and other fish markets in the latter half of 2012, employing a method similar to Bestor's (2004) inquisitive observation. I also talked to representatives of fishpen and fisherfolk associations based at the lake and in Manila, as well as state engineers and officials in agencies involved in urban flood control and the lake's environmental management. I returned to the lake villages in 2015 for follow-up interviews and also conducted additional fieldwork from 2015 to 2017 along the lakeshore sites of Muntinlupa and Taguig in southern Metro Manila and Calamba in Laguna.

For the interviews, I employed purposive and theoretical sampling, which combined snowballing and stratified purposeful techniques that targeted individuals from specific subgroups referred to me by other participants (Miles & Huberman, 1994). This strategy was adopted because of the large number of potential participants located at multiple sites and to ensure that all major fish-related livelihoods between lake and city were included in the interviews. It also facilitated access to various sites, including places that might otherwise have been difficult to enter, such as certain aquaculture production sites and the very restrictive urban fish port.

The semistructured interview schedules were composed of several questions that were subsequently adapted to fit the livelihood and initial responses of participants, inquiring about production and trading practices, access to means of production, relations with other producers, marketing and distribution of fish, place histories, and socioecological changes in the lake, among other topics. Interviews were conducted mostly in Tagalog/Filipino, the local language, coded for analytical themes and translated into English. Where appropriate in the discussion, I have included the Tagalog terms for various local names and their closest, locally used English translations.

I also spent time talking to and engaging in participant observation of everyday events at various sites at the lake to understand livelihood practices—for example, maintaining aquaculture cages, seining a fishpen, harvesting fish in a corral and cages, assembling fish for trade, strip spawning of fish in hatcheries, unloading fish—and at the city markets, fish ports, and neighborhoods where fish is consumed. Multisite research on an expansive topic such as the geographies of urban resource flows requires making analytical choices about which people, relations, or places are included in the narrative. These choices are of course a product of a partial, selective, and situated understanding of places that cannot be divorced from research positionalities. Navigating my insider/outsider position as a Filipino researcher who spoke the local language but remained an outsider to many of these communities has been shaped by particular theoretical and political commitments to understanding spatial injustices and uneven development resulting from urban metabolic relations. These orientations are embedded throughout the subsequent discussion of the various visions and practices of people as they rely on, make do with, or transform the particular socioecological configurations they continue to inhabit.

The book is divided into two parts and contains six chapters, four of which present empirical narratives that tease out urban metabolic relations between the city and the frontier. In part 1, I examine how Laguna Lake was socioecologically produced as an urban resource frontier by the intersections of state, capital, and livelihoods, generating contradictions that reconfigure resource governance and production. This half of the book narrates a history of the spatial expansion of urban resource frontiers in the lake (chapter 1), complemented by accounts of socioecological transformation in these frontiers (chapters 2 and 3). In part 2, I follow resource flows from frontier to city and back to investigate questions of access, practices, and imaginaries. I use stories of provisioning of fish and movements from spaces of production to consumption (chapters 4 and 5), and of floodwaters and infrastructure (chapter 6) to illustrate how resource flows from frontiers are encountered and transformed by everyday practices.

In chapter 1 I investigate the history of the frontier-making relations between Metro Manila and Laguna Lake throughout the second half of the twentieth century. This chapter traces imaginaries and visions of the lake as a frontier through state and scientific projects, bodies, and infrastructure that initiated, controlled, or managed fish, water, wastes, and other vital flows between the city and the lake. The modern ideology of progress and taming nature permeated the postcolonial state and underpinned visions of the subsequent developmental, authoritarian, and neoliberal modes of lake governance. I demonstrate how Laguna Lake was dreamed of and designed as a multi-use resource that would promote agrarian development while providing resource flows for Manila. Aquaculture became central to this project as a technology to produce fish more efficiently amid framings of an overexploited yet underutilized lake that rendered the lake extractable. Efforts to improve the fish, the production techniques, and the lake were necessary to realize the modern goals of development, and knowledge of the lake's nature became a significant prerequisite to control. Using the controversy over hydraulic control of saltwater flux to the lake, I also contrast the state's modern/scientific and fisherfolk's lived/practical knowledge of the lake as a form of frontier unmapping of complex lake socioecologies. The chapter shows how the lake became a modern laboratory for socioecological experimentation and new modes of production and resource governance that aimed for state managerialism of conflicting urban metabolic flows.

Chapter 2 adopts the lens of the commodity frontier to narrate the expansion and crisis of aquaculture in Laguna Lake. The emergence and entrenchment of capitalist aquaculture in the lake are rooted in changing political economies of a fishing industry that took advantage of routine failures of lake management and reshaped state regulation. The intrusion of large-scale urban investments on aquaculture and expansion of enclosures produced conflicts with small-scale fisherfolk, leading to a contentious history of accumulation, dispossession, and violence. However, periods of crisis in aquaculture production followed bursts of productivity as socioecological contradictions of production undermined accumulation. I also demonstrate how aquaculture and its associated technologies radically altered fishing village agrarian structures and relations of production. Differential access to and participation in the aquaculture economy resulted in intra- and intervillage stratification and forms of capitalist subjectification. Taking the case of two lake villages, I contrast the diverging and conjunctural fates of villagers as they become increasingly intertwined with urban-oriented aquaculture. The boom and bust cycle of aquaculture production, aquaculture's urban elite capture, state regulation, and their impacts on agrarian relations may be understood within a particular logic of how capitalism develops and reproduces through nature.

In chapter 3 I discuss three stories surrounding the problem of nature's materiality in a resource frontier. Despite modern ambitions of control and partitioning, nature's materiality often frustrates intentions and designs and takes ecological trajectories in surprising and unexpected directions. Hazards regularly undermine attempts to both control fish production through aquaculture and institute regulatory regimes to manage its excesses. Capital confronts various properties of nature in production that shape trajectories of accumulation and the relations of production in the lake. Unruly, invasive fish assemble new practices and communities as they enable various livelihood encounters. Nature's materiality emphasizes the unruly nature of frontier making in Laguna Lake, where reconfigured state, capital, and livelihood relations redefine trajectories of resource production.

Chapter 4 moves the narrative to the city by tracing the assemblage of actors, nodes, and relations that connect lake production with urban consumption. Foregrounding questions of access and social relations surrounding commodity flows, the chapter's discussion centers on how people derive benefit and exclude others from urban value chains of fish. It explores everyday economic transactions and social ties that structure urban flows of fish, identifying elite power and mechanisms of exclusion at various stages of the chain. The chapter's narrative travels from fish nurseries and farms in the lake to fish ports' practices of fish exchange and labor relations in Manila. Combining political economic access analysis with cultural economic commodity lenses, I map out who benefits and who is excluded from flows of urban provisioning and how.

Chapter 5 follows the story of bighead carp to show the material transformations of fish from the lake to city. It uncovers the various social relations and urban practices of transforming fish to overcome inherent contradictions of fish production for the city. The chapter demonstrates the significance of Laguna Lake fish flows for Manila's fish consumption and shows the work undertaken as fish travel from sites of production to those of consumption. Focusing on the "biographies" of bighead carp from the lake, I present narratives of formal and informal, as well as cultural and economic, practices as people transform bodies and imaginaries of fish through distancing and entanglement. I show how relations and practices of people transform flows of fish between lake and the city as a different way of illustrating what constitutes urban metabolism.

Chapter 6 parallels the stories of fish flows in the previous chapter and tracks the history and politics of how floodwaters are transferred, displaced, and managed. The devastating floods of 2009 and 2012 in Manila made visible the breaking down of modern infrastructure networks that underpinned urban stormwater flows. Yet urban flood infrastructure networks have a deep history that connects with Laguna Lake as a necessary component in their functioning. The infrastructural design sought to solve flooding in Manila by transferring risk elsewhere. This chapter tracks the material and ideological construction of Laguna Lake as a hidden landscape where urban risk could take place. Living with magnified flood risk in turn reconfigures the everyday lives of lake dwellers and their livelihood possibilities. Through a focus on infrastructure, the chapter demonstrates how urban risk embodies the relations between city and frontier and how state control of flows is predicated on imaginaries of space and the work of infrastructural maintenance, with significant implications for lake lives.

Finally, the epilogue briefly wraps up the book's narrative by posing questions about the socioecological futures and the spaces of politics presented by an urban metabolic lens on a changing frontier. Recent infrastructural controversies and neoliberal/populist shifts in governance of Laguna Lake bring the politics, temporality, and spatiality of urban metabolism to the fore again. Amid the politics of ruination and death of the lake, the epilogue ends with possibilities for remaking socionatures.

PART ONE

Making and Remaking a Frontier

ONE

Birth of a Convenient Frontier

FICTION AND HISTORY have long documented imaginations of Laguna Lake as both a problem and a resource. Jose Rizal opens *El Filibusterismo* (*The Subversive*), his satirical novel about late nineteenth-century colonial Manila society, with an image of the white steamship *Tabo* navigating upstream along the Pasig River toward Laguna Lake.[1] The ship sails slowly as skippers and sailors attempt to negotiate the river's meanders, its shallow waters, and the stretch of sandbars at the river's mouth where it meets the lake. While most passengers crowd below the ship's deck, the European elites and colonial officials sit shaded on the deck, gazing at the still waters as they debate how best to solve the problem posed by the silting river and the shallow lake. Among the proposals floated are using forced labor to dig a stream channel through the city and encouraging people to dig in the sandbars to gather snails as feed for the flourishing local industry of duck raising and the production of *balut* or fertilized duck embryo. The forced labor proposal brings up questions of whether the people might revolt against the Spaniards again, while the snail suggestion elicits a snobbish retort from one of the deck's passengers: "If everyone were to breed ducks there would be an excess of *balut* eggs. Ugh! How disgusting! Leave the sandbars alone!" (Rizal, 2007, p. 10).

A parallel image opens Ishmael Bernal's (1976) film, *Nunal sa Tubig* (*Speck in the Water*).[2] From a shot of smoke rising from a factory stack, the camera pans to a group of men dressed in white cruising on a speedboat. A white man and his local business partner gaze at the landscape as they speed across the calm lake waters, where scenes of traditional rural life continue: a slow passenger boat packed with people, a pair of women catching mudfish underneath clumps of water hyacinths, and men mending nets. Their boat stops in one of the aquaculture enclosures shown to have taken over the lake, as if to survey the lake as a potential business venture. The men nod in satisfaction, then they are interrupted by a threatening sky that forces them to return to shore. The camera then cuts to the vantage of villagers, many of whom scurry home as the winds pick up, to take shelter before an impending downpour. The harbinger of an approaching storm and the cut to disrupted village life signals the shift in the film's focus from the nameless entrepreneurs searching for a business opportunity to the stories and struggles of villagers amid rapid social and environmental change in Laguna Lake in the 1970s.

Rizal's first chapter alludes to state power, elite interests, and the work required in navigating and remaking lake and urban environments. The debates between the characters of the ship's upper deck present a prelude to real twentieth-century state infrastructure projects that asserted a sense of control, even if illusory, over unruly and uncooperative lake nature intimately connected to the city. These massive projects attempted to solve the problem that the landscape presented and relied on an assemblage of techniques and imaginaries that simplified, effaced, and instituted novel arrangements while framing the lake as a problematic frontier ripe with opportunities. Bernal's surveying entrepreneurs, meanwhile, represent the early days of the opening up of the lake as a frontier for commodification, when a scramble to build profitable aquaculture enclosures dispossessed and marginalized lake dwellers. The contrast between the men in speedboats and the timelessness of rural life in the villages reflects the conflicting imaginaries that would permeate early Laguna Lake development plans: the inevitable march of progress needed to replace the slower, traditional ways. The foreboding storm presciently anticipates the radical socioecological changes that accompanied aquaculture expansion and industrialization just a few years after the film's release, which shaped the struggles of villagers living in a transformed environment.

Both fictional accounts mirror historical interventions by state and capital that sought to make Laguna Lake perform the work of a resource frontier. In the early 1960s, barely a decade after the end of nearly half a century of American colonial rule, the nascent Philippine state turned to Laguna Lake as part of its quest to spur national development. When Senator Wenceslao Lagumbay, a key proponent of legislation regarding Laguna Lake, stated in 1966 that "the development of Laguna Lake will make a dream come true," he was echoing narratives of optimism of his time (Caliwag, 1966, p. 27). Politicians and statesmen were eager to transform the lake from its traditional subsistence fishing–based economy into a multi-use resource space attuned to the emerging needs of both rural and urban populations. These desires were situated within emerging imaginaries of national development and modern resource control that suffused postindependence state building.

This chapter traces the early periods of making Laguna Lake Manila's resource frontier through state projects designed to extract and deliver vital resource flows. Frontiers have always been seen as spaces to be transformed to enable exploitation, drawing margins closer to the state's territorializing power (Cons & Eilenberg, 2019). Imaginaries and new magical visions of place accompany their birth, conjured as sites of potential and desire while simultaneously marked as underdeveloped, unproductive, and empty (Li, 2014; Tsing, 2005). Laguna Lake has been imagined as a multi-use resource and framed as overexploited but also underutilized. These creative visions necessarily seek to erase existing knowledge relations through unmapping practices of simplification and legibility and by the introduction of modern solutions for a problematic nature that paradoxically result in newer ecological problems.

Elsewhere in the Philippines and throughout Southeast Asia, mid-twentieth-century frontier making was a history of vast lands at the margins being exploited and brought into the orbit of the state for resource extraction, often through the exercise of power and displays of violence (Cons & Eilenberg, 2019; Dressler & Guieb, 2015; Nevins & Peluso, 2008; Tsing, 2005). Laguna Lake is a distinct resource frontier in this regard as an *urban* resource frontier where techniques of rural development intersected with desires to provision the city. However, the history of frontier making shows similar processes of simplifying spaces and socioecologies and silencing other spatial and socioecological narratives that do not fit within state imaginaries of development and that subsequently haunt these very visions of modernity. The lake is imbricated in various scales of state political ecology tied to urbanization, modernity, and technoscientific knowledge of nature.

Laguna Lake has always been an important source of fish and crustaceans for its immediate surrounding region, with the opening scene of Rizal's novel attesting to a long history of urban connections facilitated by the lake's fluidity and productivity. The lake supplies the city with food and serves as a node in a water transport network that connects to the sea. Yet the lake environment has been seen as riddled with problems that have justified more state interventions, from its shallowing that made navigation difficult to its uniquely eutrophic character that has not been efficiently utilized. As a space of potentials, the lake served as a site for the postcolonial state recovering from the damage of war to flex its ambitious vision for development.

Spurring Laguna Lake development as a resource, politicians would repeat over the years, was a dream waiting to be realized.

DEVELOPMENTAL DREAMS AND AUTHORITARIAN ANXIETIES

On July 18, 1966, the Philippine Congress passed Republic Act No. 4850, creating the Laguna Lake Development Authority (LLDA). In many ways, the LLDA was a unique and pioneering political body in the country: a self-sustaining, quasi-state corporation that aimed to develop and regulate Laguna Lake's resources. Its jurisdiction covered an area on the scale of the watershed—including the whole Laguna Lake and the rivers that drain into it—and it became a precedent for other regional development projects on the basin scale (see figure 1 in the introduction).

Laguna Lake's development was a dream waiting to be realized through modern principles of control of nature: "Control of the lake is an indispensable element for the proper physical planning and development" (Laguna Lake Development Authority, 1966, p. 2), as stated in a prospectus. The language of modern control and productivity permeated calls for efficient planning to take advantage of "the rich but untapped resources of the Laguna Lake Area," which had yet to "be harnessed by a fully organized, long-range development strategy into an effective development event that will yield the best results" (Laguna Lake Development Authority, 1966, p. 6).

The "grand vision" and ambitious development plan for the lake involved agricultural, industrial, and tourism growth facilitated by reordering of the lake's nature and its management. Reclamation, dikes, flood control, aquaculture, and other infrastructure and social development projects would eventually emerge, spanning two provinces and Metropolitan Manila, an administrative urban region that formed amid consolidated power under authoritarian rule in the 1970s. Technocrats, politicians, fishery scientists, engineers, foreign consultants, and military officials would populate the new bodies created to manage problems of this new space of resource management. Owing to its liminal location on the edge, Laguna Lake's resource frontier making was framed as both a rural and an urban development problem.

The LLDA began as a modest organization that got off to a shaky start. It was understaffed and often ran into conflict within its ranks and with other local units under its jurisdiction (Cruz, 1982; Delmendo & Rabanal, 1982).

Despite grand and ambitious modern goals of producing the lake as a multi-use resource set out in the 1966 legislation, it took a few more years before the LLDA secured financing and was formally organized on May 30, 1969, at the home of Senator Helena Z. Benitez (Florendo, 1969). It launched its first project a year later in 1970: the introduction of a new and more efficient way of producing fish through aquaculture.

Yet aquaculture was not the primary nor the most important developmental project of the LLDA, despite subsequent events that placed it centrally in resource management in the last half of the century. The year that the LLDA was signed into law, the Philippine government requested assistance from the United Nations Development Programme (UNDP) to conduct scientific studies to assess the developmental potential of the lake as a water resource (Florendo, 1969; Laguna Lake Development Authority, 1966). Laguna Lake resource production had also been framed as dual-purpose control of water for the city: to help rid Manila of floods while supplying domestic water to meet future urban demand.

The immediate precursor of the LLDA's efforts to manage metabolic flows of water was President Diosdado Macapagal's (1961–1965) Program Implementation Agency, which sought to solve Manila's persistent flooding through control of Laguna Lake's overflow (Florendo, 1969; Macatuno, 1966). As with previous postwar plans of the state however, this project did not materialize, for a variety of political, financial, and technical reasons (see chapter 6). Macapagal's successor, Ferdinand Marcos (1965–1986), expanded the initial plans from flood control to a broader development project (Macatuno, 1966). The high modern ambitions of development through control of nature promised by the LLDA's creation would find a match in Marcos's turn to centralizing authoritarianism. Declaring martial law in 1972, he sought to quell the threat of communist rebellion by consolidating oligarchic and military power through a conjugal dictatorship with First Lady Imelda Marcos (Mijares, 2017). Laguna Lake's development trajectories would bear the imprints of the triple technocratic developmental features of Marcos's authoritarian regime: the introduction of green revolution technologies, the promotion of export agriculture, and reliance on foreign borrowing (Boyce, 1993; Ofreneo, 1980).

By instituting infrastructural, livelihood, and regulatory interventions to address developmental problems, Laguna Lake became an arena for Marcos's vision of development, playing a role in his desire to create a "New Society" out of the destruction of the old political and social order.[3] The lake was a site of rural unrest during the martial law years (1972–1981), and Marcos's New

Society sought to suppress dissent while promoting development through modern projects. Marcos secured foreign loans to fund his large, grand infrastructure projects in Manila and elsewhere. The infrastructural obsession with grand structures—termed by Lico (2003) an "edifice complex"—was to be a showcase of authoritarian power and a display of the radical change promised by the New Society.

In Laguna Lake, this infrastructural obsession gave birth to a hydraulic control structure and several flood control structures, which along with an aquaculture development project and a cooperative development project were the major developmental interventions in the lake during the martial law years. Yet as high modern symbols of authoritarian excess, these projects also embody their contradictions and failures. Fish producers opposed the hydraulic control structure and were eventually successful in shutting down its operations. Flood control structures that seemed to have kept the city dry for a while broke down decades later and caused unprecedented disastrous damage. The aquaculture and cooperative development projects, despite millions of pesos in investments, were massive social development failures.

The martial law years under Marcos were also marked by rapid urbanization and a more explicit state recognition of urban problems. Fueled by migration from a troubled countryside, Metro Manila's population more than doubled between 1950 and 1970 and again between 1970 and 1990, with many of these "surplus populations" finding homes in slums located on marginal and hazardous lands and often subject to state eviction (Karaos, 1993). The problem of feeding Manila, while historically nothing new (Doeppers, 2016), became quantitatively magnified during this time of rapid urban growth. The increasing pressure to produce Laguna Lake as a resource became more than a matter of rural development and was subsequently aligned with finding solutions to urban questions. The resource work that Laguna Lake was expected to perform—as a space to store stormwater, as a source of urban domestic water, and as a steady supplier of fish—was to be made possible by application of modern techniques of infrastructure and resource management.

Marcos was ousted through a people power revolution in 1986. Subsequent administrations adopted a variety of liberal and neoliberal modes of governance of the lake and resource production in general. However, the legacy of authoritarian rule and infrastructural interventions remains visible in the landscape and in the regulatory strategies for managing the lake's resource problems. Perhaps no other intervention has radically altered lake socioecologies more than aquaculture.

EXPERIMENTING ON THE FRONTIER: INTRODUCING AQUACULTURE

Aquaculture through fishpen and fishcage technologies presented a radically new way of producing fish, one without precedent in Laguna Lake or the Philippines. Whereas pond aquaculture, first documented in the Visayas in the fifteenth century, evolved slowly over time (Schmittou et al., 1983; Villaluz, 1950), pen and cage culture in large water bodies represents late twentieth-century technologies that very rapidly transformed their host spaces. As a means of producing aquatic organisms, it differs fundamentally from capture fisheries in property rights and ownership, degree of control of production, and potential for further growth (Belton & Thilsted, 2014; Campling & Havice, 2014).

In contrast to capture fisheries, where fish are generally not owned until caught, property rights in aquaculture are clearly assigned to the producer from seed to harvest stage. Aquaculture also involves closer control and management of the production process. While extensive and low-intensity aquaculture has ancient origins, several other forms of aquaculture technologies have developed based on production environments (marine, brackish water, and freshwater), production types (pond, cage, net pen), and production intensification (intensive, semi-intensive, and extensive). Aquaculture has always been seen to have brighter prospects for future global growth than capture fisheries as a result of its dramatic rise over the last half century, with global production ballooning from 1 million metric tons in 1950 to 80 million in 2016 (FAO, 2018).

Aquaculture emerged as the first of several state-improvement efforts designed to gain greater control of lake nature. By the 1970s, Laguna Lake had become a pioneering site for pen and cage aquaculture technologies. Fishpens are extensive enclosures (ranging in size from one to hundreds of hectares) with structures made of synthetic mesh nets tied to bamboo poles staked to the muddy bottom of the lake. Fishcages are much smaller, stationary structures, often less than a hectare in size, made of inverted mosquito nets that may or may not be covered, which unlike fishpens do not make contact with the lake bottom. Aquaculture subsequently altered lake fisheries production and catch composition for decades to come (see figure 2).

Experimental efforts to develop aquaculture technologies in the lake were always situated within the goal of introducing these technologies to water bodies elsewhere in the country. More than a decade after they were

FIGURE 2. A fishpen enclosure in the middle of Laguna Lake, 2012. Photo by author.

introduced, the head of the Bureau of Fisheries summarized aquaculture's transformative impacts in improving lake productivity: "When fishpens were introduced in Laguna de Bay and have been proven viable, the lake was transformed from a near barren lake to a productive one which not only revolutionized aquaculture but also made the lake the country's showcase in aquaculture technology" (Mane, 1983, p. 31).

The framing of a "near barren lake" fits within the context of crisis and decline in capture fisheries in the lake. The introduction of motorized fishing gear in the 1950s led to overexploitation of snails and shrimp that served as feed for a flourishing duck industry along the lake's shore, famous for producing *balut* (the same lakeside food that one of Rizal's characters found disgusting).[4] Crustaceans and snails comprised as much as 70 percent of the total capture fisheries catch of 350,000 metric tons (MT) in the lake in 1963 (Davies et al., 1986; Rabanal et al., 1968, Richter, 2001). Already in significant decline, by the 1960s the snail fisheries had almost collapsed.

Capture fisheries have a long history in Laguna Lake and have diversified through the use of a variety of gear (see table 1). Motorized push nets (*suro* or *sakag*) are a nonselective type of gear that can capture fish of all types and sizes. Up to five times more efficient than other gear, this fishing technique was introduced in the 1950s and employed fisherfolk in several villages for the next few decades. Drag seines (*pukot*) were the dominant fishing gear used at the turn of the twentieth century, as noted by early studies of lake

TABLE 1 Common Capture Fisheries Gear in Laguna Lake

Category	Motorized Push Nets (*Sakag, Suro*)	Drag Seine (*Pukot*)	Gill Net (*Pante*)	Drift Long Line (*Kitang*)	Cast Net (*Dala*)
Catch per unit effort (kg/hr. 1995–1997)[a]	3.64	0.65	0.50	0.70	0.70
Species caught[a,b]	Nonselective: all kinds, including shrimp and fish fry	Selective: silver perch, white goby, Manila catfish, mudfish	Selective: tilapia, bighead carp, Manila catfish, milkfish	Selective: knife fish, white goby	Selective: mudfish, catfish
Labor needs (persons/boat)[b]	5–12	10–16	2–4	2–4	2–4

[a]From Palma et al. (2005).
[b]From author's fieldwork.

fisheries (Aldaba, 1931b, 1931c; Mane & Villaluz, 1939; Villadolid, 1934). Both motorized push nets and drag seines, owned by wealthier village households, employed a crew of more than a dozen, led by captains or drivers. Gillnets (*pante*), long lines (*kitang*), and cast nets (*dala*) are active gear organized in much smaller boats and have lower catch per unit effort when compared with motorized push nets.

A brief note on the terms *fisherfolk* and *capture fisherfolk* is necessary. Fisherfolk is a collective term that encompasses various livelihood activities and identities surrounding fishing and extraction of lake fauna. The local terms *mamamalakaya* and *mangingisda* (literally translated as "someone who engages in fishing") reflect how these identities are defined by their engagement with fishing as an activity in the lake. Fishing includes the use of a variety of techniques, gear, relations, and involvement in the realm of production. Fisherfolk further refers to subidentities based on these differences: *suro* and *sakag* users are referred to as *manunuro* and *mananakag*, *pukot* fishers as *mamumukot*, *pante* fishers as *mamamante*, and so on.

Fisherfolk as a collective term is also used loosely to include many other associated fishing activities beyond direct extraction practiced by both men and women, such as vending, drying, processing, assembling, trading, and aquaculture seining/harvesting, as well as the use of passive fishing gear such as fish traps and corrals. Because it is defined by a specific activity, belonging is dynamic and nonexclusive, meaning one could engage in part-time work beyond fishing or completely shift to other livelihoods (such as aquaculture or migrant work in the city) and still adhere to a fisherfolk identity. This complicates simplistic distinctions between small-scale aquaculture producers, many of whom were also previously engaged in full-time fishing or are still involved in certain fishing activities. The implications of how small-scale aquaculture producers position themselves in relation to long-standing aquaculture-fisherfolk conflicts is explored further in chapter 2.

Laguna Lake has long supplied freshwater fish for local and regional consumption, including indigenous species such as silver perch or *ayungin* (*Therapon plumbeus*), white goby or *biya* (*Glossogobius giurus*), and Manila catfish or *kanduli* (*Arius manilensis*). However, finfishes for human consumption only utilized 7 percent of the primary production in the lake. For the state and scientists, this low utilization suggested a vast potential for converting abundant abiotic nutrients to fish flesh and human food through naturally occurring plankton (Delos Reyes, 1993; Laguna Lake Development Authority, 1970, 1978b).

Fishery scientists and the LLDA argued that capture fisheries alone could not meet the goal of producing enough fresh fish to meet demand because they only maintained "fishing families on subsistence level of livelihood and failed to tap the lake's potential as a major fishery resource" (Laguna Lake Development Authority, 1982, p. 5). This dual framing of the lake as both overexploited and underutilized provided the impetus to develop a more modern, more efficient, and more productive way of producing fish through aquaculture. Such plans relied on a mixture of external funding, fisheries experimentation, and novel understanding of the lake's unique ecology.

Following recommendations from a series of studies funded by the UNDP and with the goal of determining aquaculture's feasibility, in 1970 the LLDA established an experimental farm in the lake's Central Bay (Laguna Lake Development Authority, 1978b). The 38-hectare Looc Fish Pen Demonstration Project, constructed in the town of Cardona, Rizal, stocked milkfish or *bangus*, a non-native species chosen for its high market price and consumer desirability and because it fed primarily on the lake's phytoplankton (Delmendo & Gedney, 1976). Milkfish has historically been a daily staple for Manila residents, according to a survey the most consumed fish in the city in the 1920s and 1930s and perhaps even decades earlier (Doeppers, 2016; Herre & Mendoza, 1929). The fish has traditionally been farmed commercially in fishponds around old Manila, including Malabon, Navotas, Bulacan, Pampanga, and Pangasinan, where proximity to urban markets made it a lucrative venture.[5] Wild milkfish had found their way into the lake previously but became increasingly rare by the 1920s.[6] The experimental farm introduced the fish in commercial and controlled quantities.

The experimental project used the novel fishpen technology (see figure 2). Pen culture methods were first documented outside the Philippines in Japan in the 1920s and in China in the 1950s (Beveridge, 1984). In Laguna Lake, the LLDA designed the earliest pens as larger versions of the catching chamber of *baklad* or fish corrals, small-scale structures that had existed in the lake and in other water bodies such as Manila Bay, which were used by fishers to trap fish passively (Delmendo & Rabanal, 1982; Doeppers, 2016).

As a reflection of the project's importance to the state's development agenda, President Marcos sent his executive secretary to grace the ceremonial first harvest on July 9, 1971 (Laguna Lake Development Authority, 1971). The harvest demonstrated that pen aquaculture was feasible in the lake and supported earlier assessments of the potential for increased efficiency and yields in fish production. Five months of culturing milkfish without any artificial feeds yielded 700 kg per hectare, and the LLDA estimated the possibility of an annual yield of 1,500 kg per hectare, a figure more than three times that of capture fisheries (Delmendo & Gedney, 1976). By the early 1980s, the LLDA would record dramatically high productivity of up to 10,000 to 15,000 kg per hectare, as opposed to the 330 kg per hectare per year in capture fisheries (*Bulletin Today*, 1984d). The lake had been so highly productive by then that it supplied 80 percent of the freshwater fish needs of Metro Manila and 30 percent of the country's (Mane, 1983).

In the fishpens, LLDA aquaculturists also experimented with a different production technique. Fishcages were introduced as part of the Floating Cage Project as smaller-scale alternatives to fishpens. The feasibility of raising carp, tilapia, and milkfish was tested in cages from 1973 until the 1980s. Nile tilapia displayed the best growth among these species, growing to a marketable size in four months (Garcia & Medina, 1987; Guerrero, 1981; Laguna Lake Development Authority, 1974). Cages were suited for tilapia, which displayed the behavior of burrowing in the mud that made harvesting them from pens difficult. Because the fishcages were much smaller and easier to manage, the project eased fisherfolk access to small-scale aquaculture. Both large-scale fishpens and small-scale fishcages continue to coexist in the lake as a result of these projects (see chapter 2). Milkfish, tilapia, and bighead carp

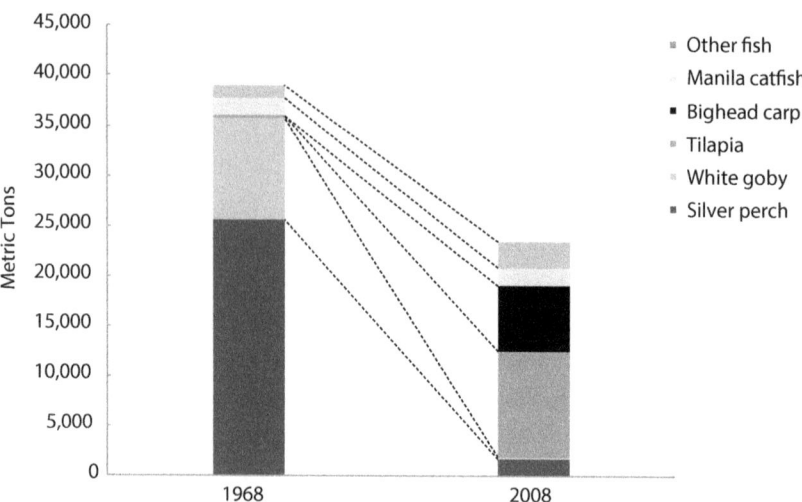

FIGURE 3. Laguna Lake capture fisheries catch composition, 1968 and 2008. *Sources:* Mercene (1987); Philippine Statistics Authority OpenSTAT database.

(introduced more recently) eventually came to dominate fish catch composition of capture fisheries over the next decades (see figure 3).

From its origin, the LLDA had in view the goal of establishing commercial-scale aquaculture operations, including an expansion to 20,000 hectares through an externally funded fishpen development project (Delmendo & Gedney, 1976; Laguna Lake Development Authority, 1970, 1972). During the ceremonial first harvest at the experimental farm in 1971, an LLDA official reiterated the intention to transfer fishpen technical knowledge to fisherfolk (Cruz, 1982). However, delays in instituting financial assistance mechanisms for fishers, the attractiveness and profitability of pen operations, and regulatory conflicts with local governments in issuing pen permits contributed to the rapid entry of well-capitalized entrepreneurs at the expense of fisherfolk involvement in fishpen aquaculture (Delmendo & Rabanal, 1982). Thus, despite successfully proving the technical feasibility of growing fish in the lake in pens and cages, the LLDA failed to recognize the social contexts of fisherfolk livelihoods, the political and economic power of urban elites and middle-class entrepreneurs, and the complexities of lake governance.

Aquaculture promised efficient and controlled fish production to improve livelihoods and food security, which overexploited and underutilized capture fisheries in the lake were deemed unable to provide. Traditional capture fisheries produced a diverse variety of indigenous fish, but these were small in size and fetched low prices in the urban market. Quantity and volume of fish related to value and ability to convert natural feeds into protein became a more pressing concern than maintaining a healthy diversity of fish for fisherfolk subsistence. Enabling the growth of aquaculture required the reworking of fish bodies, techniques of production, and the lake environment. Experimental farms, research stations, and scientists contributed to the first two needs, while infrastructure and governance strategies were necessary in the more complicated third task.

INNOVATING ON THE EARLY FRONTIER: IMPROVING AQUACULTURE

The ideal of a productive lake relied on efficient production of fish. The introduction of nonindigenous but potentially valuable fish species such as milkfish and various experiments with techniques were both crucial components in achieving this goal. Due to limited expertise in fisheries science, for aquaculture improvement work the LLDA partnered with other scientific institutions such as the Southeast Asian Fisheries Development Center (SEAFDEC), a regional intergovernmental fisheries development organization, which established an inland freshwater station in the lake in 1976 (Delmendo & Rabanal, 1982; Platon, 2001). The station played a prominent role in the development of lake aquaculture through its research on and dissemination of fish seeds to villages.

SEAFDEC and the LLDA tested various species such as tilapia and carp in a variety of production scenarios. For example, the Mozambique tilapia (*Oreochromis mossambicus*) had been introduced in the lake in the 1950s, but its growth and market desirability were found to be low (Asian Development Bank, 2005). The bigger and tastier Nile tilapia replaced the Mozambique species, although its behavior of burrowing under the mud during harvest and its ability to escape fishpens prompted further innovation in production techniques. Studies found floating cages to be more appropriate for the fish's behavior and tolerance for higher stocking densities (Richter, 2001). Research under the Polyculture Development Program in the mid-1980s also suggested better tilapia growth in cages when paired with other fish like bighead carp, leading to the eventual expansion of the technique in the lake (Gonzales, 1984; Tabbu et al., 1986). The program aimed to improve productivity through efficient use of constrained space amid emerging conflicts in the lake. The LLDA's general manager promoted polyculture in early 1984 as an efficient use of lake space as it contributed to making "all levels of Laguna Lake productive without eating habits of finfishes, crustaceans and molluscs at different levels interfering with one another'" (Saquin, 1984, p. 37).

The experimental farm and SEAFDEC-LLDA studies also demonstrated that planktivore fish such as milkfish and tilapia can grow to marketable size without supplemental feeding, and they examined the (im)practicality of formulated feed use (Platon, 2001; Richter, 2001). Laguna Lake's highly eutrophic, nutrient-rich character enables the possibility of extensive and semi-intensive aquaculture without or with only minimal artificial feed. SEAFDEC's tilapia and bighead carp research enabled the widespread production of the two species in the lake (Basiao, 1994; Bautista et al., 1988; Fermin, 1991; Gonzales, 1984; Romana-Eguia & Doyle 1992; Santiago et al., 1988). Its tilapia research complemented other projects elsewhere that aimed to improve tilapia production, such as the ADB- and UNDP-funded Genetically Improved Farmed Tilapia (GIFT), which used selective breeding methods to improve Nile tilapia strains that since 1988 had been distributed throughout the world and were responsible for higher tilapia yields (Eknath & Acosta, 1998).[7] It is uncertain to what extent the tilapia strains in Laguna Lake have benefited from such projects, given the extensive improvisation and informal hybridization of strains in seed production, but these technologies have certainly contributed to larger and faster-growing tilapia strains in the lake.

These early projects showed how attempts to produce bigger and better fish for human consumption involved alterations of production techniques, fish behavior, and the nature of the fish itself. In the 1980s, yield increases of up to three or four times were made possible in part by the choice of the best-suited fish that would make the most efficient use of space (Saquin, 1984). However, producing better fish alone was not enough to ensure higher productivity. Laguna Lake fisheries depended heavily on the quality of the lake's water. The state recognized knowledge and control of the water and lake environment as crucial to the success of fish productivity and lake development.

EXPERIMENTAL ECOLOGIES: IMPROVING THE LAKE

> By preventing pollution and the intrusion of salt water, the lake can be developed as a source of water supply for the communities along the lake shore; the duck raising industry will be regenerated;

> municipal fishing within the lake will be improved; and the lake can be made a source of water for irrigation. In addition, control of the lake is an important factor in solving the sewerage problem of the City of Manila through the reduction of flooding
>
> LAGUNA LAKE DEVELOPMENT AUTHORITY *(1966, p. 2)*

Perhaps no other statement better captures the faith in the modern ideal of control of lake processes to initiate a whole host of developmental interventions and extract multiple benefits from a resource. Laguna Lake is a good example of how the control of the environment is necessary to achieve improved fish production and to enable various other aspects of development of a resource. Laguna Lake is unique in its eutrophic character, which encourages growth of phytoplankton, creating the conditions for highly productive fisheries. With an average depth of 2.5 meters, it is also shallow as a result of siltation from surrounding activities, which contributes to its seasonal turbidity that in turn constrains fish growth. The saline flux from the sea through the Pasig River and Manila Bay regularly improves water condition by helping reduce turbidity and encouraging plankton production (see map 2).

In attempts to make the lake perform its multiresource work, the complex socionatures and spatialities that produce the lake—biotic/abiotic and urban/rural—were necessarily simplified to introduce and justify modern interventions. Infrastructure was crucial in this picture, as "the key point to the overall Laguna de Bay development scheme," which becomes "indispensable to emphasize or support the other various projects for water supply, irrigation, fishery industry, etc." (Pacific Consultants International, 1978, p. 6). The controversy over regulation of saltwater intrusion into the lake through hydraulic control as an ecological simplification for infrastructure intervention is a case in point.

Scientific work on the unique limnological characteristics of the lake had been undertaken as early as the colonial-era 1930s by American-trained Filipino fisheries scientists. Early studies described the state of fish production in the lake, identified threats to the fisheries, and recommended appropriate regulatory action (Aldaba, 1931b; Mane & Villaluz, 1939; Villadolid, 1933). Cendana and Mane's (1937) article, for instance, had already identified the seasonal saline backflow from the Pasig River and its contradictory effects on the fisheries through observations of chlorine content in various sample stations in the lake. These recordings were supported by observation of

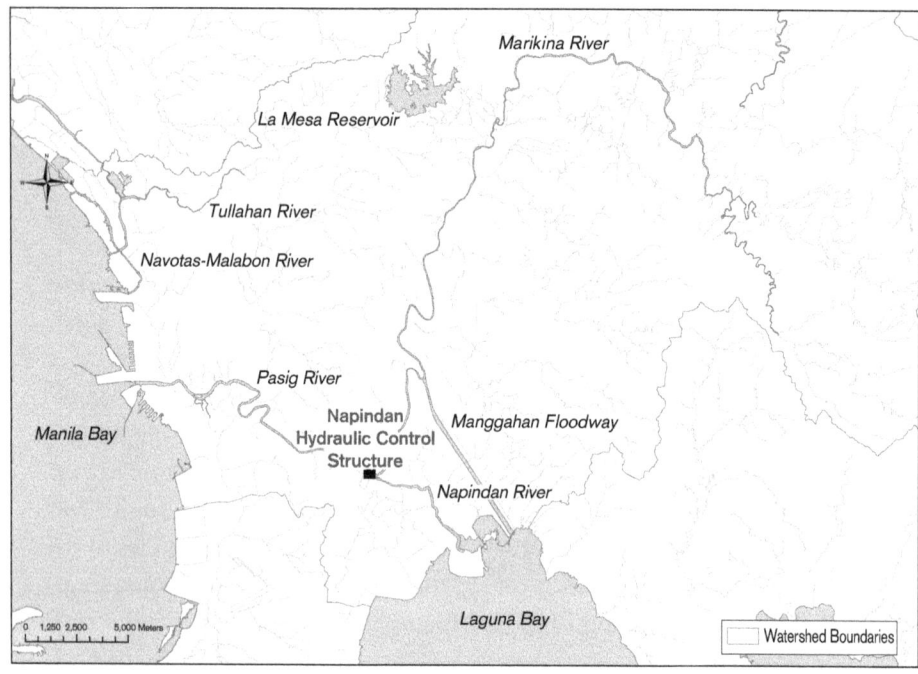

MAP 2. Waterways and flood infrastructure in the Laguna Lake Basin. Map by Patricia Anne Delmendo.

fishers whose fishing behavior adjusted with the spatiotemporal variation of diffusion of salinity in the lake. They noted that "invasion of the lake by salt water resulted in unusual transparency in the lake water" and that "with the increase in salinity of the lake water, an increase in relative abundance of certain forms of animal life in the lake and apparent decrease in others were observed" (Cendana & Mane, 1937, p. 335).

Beginning in 1966, as part of its efforts to initiate development in the lake, the LLDA commissioned several externally funded studies that assessed the lake's resource potential. These studies were financed or supported by the United Nations (UN), the Asian Development Bank (ADB), the World Bank, the United States Agency for International Development, and the World Health Organization. Foreign experts and consultants collected various limnological data and recommended appropriate interventions based on problems that were identified as constraining development. Water management—for public water supply, irrigation, fisheries, and flood control—was considered essential in the overall developmental blueprint of the lake resource.

One of the most contentious issues was whether to allow continued seasonal saltwater intrusion into the lake. The Société Grenobloise d'Etudes et d'Application Hydrauliques (SOGREAH), a French consulting firm hired by the ADB, undertook limnological studies in 1972–1974 that focused on understanding the unique hydraulic regime of the Laguna Lake-Manila complex. SOGREAH supported an earlier study that recommended the construction of a hydraulic control structure to regulate Pasig River backflow into the lake (United Nations, 1970). The structure was necessary to optimally realize the lake's potential for economic production by controlling saltwater flux that accompanied the seasonal flow from Manila Bay through the river (Laguna Lake Development Authority, 1978b; Rey, 1987; Santos-Borja, 1994; SOGREAH, 1991).

The SOGREAH consultants identified backflow from the Pasig River as the biggest threat to the lake as a source of fish, drinking water, and irrigation. This saltwater flow brought excessive nitrogen flux, which they observed was responsible for episodes of *Microcystis* algae blooms that caused massive fish mortality in pens in 1972 and 1973 (Laguna Lake Development Authority, 1978b; National Statistical Coordinating Board, 1999; Santiago, 1993; SOGREAH, 1991). The saltwater intrusion limited the lake's potential for further use as a source of potable water and irrigation. Thus, the need to control this backflow became a priority program, which the Public Works Ministry realized with the completion of the Napindan Hydraulic Control Structure (NHCS) in 1983. Control of nutrient flux from the Pasig River was seen as necessary to maintain a healthy fishery while enabling further prospects for the lake as a multi-use water resource.

The NHCS was constructed between 1977 and 1982 with financial assistance from the ADB. The gated dam and navigation locks, constructed in Metro Manila near the confluence of the Pasig-Napindan-Marikina channels (see map 2), were designed to control lake level, quality of lake water, and flooding. Only a few months into its operation, however, its gates were forced to keep flow open due to unified opposition from both aquaculture producers and capture fishers in the lake. Because of its importance in production, fisherfolk often describe saltwater intrusion using various metaphors: as nutrient-giving (in the corporal sense), as a wound-healing flow that gives life and spice to a fresh body of water, and as the lake's "multivitamins." They also view saltwater intrusion as natural and the state's infrastructural interventions as responsible for its loss. Fish producers believed that the gate and its control of the saltwater backflow led to decreased fish productivity and increased fish kills.

Opposition to the hydraulic structure was also shared by fisheries experts who were not part of the LLDA. This conflict in expert ecological knowledge came to a head in a forum in 1981. SEAFDEC experts expressed their concern that instead of improving water conditions, the NHCS would achieve a contrary effect, increasing the fish kills in the lake. SOGREAH consultants, they argued, had based their recommendations on the faulty assumption that nitrogen was the primary limiting factor for plankton growth. The SEAFDEC experts pointed to inorganic turbidity instead as responsible for mass fish mortalities in the lake. Saltwater intrusion decreased turbidity in the lake (as observed earlier by Cendana and Mane), which came with the backflow alongside pollution and other nutrients that the hydraulic control structure sought to regulate. LLDA technical officials, in response, reiterated the expected multiple benefits of the project in its defense, citing the projected benefits for the different sectors: P900 million for fisheries, P27.4 billion for water supply, and P3.2 billion for irrigation. For the LLDA officials, projections of water extraction—about 3.5 billion cubic meters—to meet the future water demands of Metro Manila and surrounding regions were a strong enough justification for the project (Bandayrel, 1981).

The saltwater intrusion controversy demonstrated how technical assistance through feasibility studies and scientific assessments laid the foundation for the production of knowledge about the lake, but such assessments were made within the limnological contexts and experiences of foreign consultants, specifically of western European lakes (Santiago, 1993). Observations by prewar fishery scientists and fisherfolk's ecological knowledge about the contradictions of saline backflow as both a blessing and a curse were overlooked in favor of a simplifying discourse that identified the flow as an ecological problem—with nitrogen as the limiting factor—that needed to be overcome to promote the lake's resource development. Elimination of the backflow through hydraulic control potentially benefited the lake as a source of public water supply and irrigation but threatened lake fisheries, whose productivity relied on the saltwater's ability to reduce turbidity, improve photosynthetic activity, and increase plankton production. The accounting of potential benefits through economic valuation further justified the necessary simplification of complex socioecologies through control of flows via infrastructure.

The NHCS issue embodied in many ways the contradictions of modern visions of a multiple-use resource designed to meet the needs of an urbanizing region. In an effort to produce the lake as a multipurpose resource and make it legible for developmental intervention, state and science smoothed out and simplified spatiotemporally uneven processes to make them amenable to technical and infrastructural control. As a form of frontier unmapping, this abstract, narrow, and utilitarian state vision contrasts with and silences the practical knowledge of fish producers who live with the complexity of lake nature through everyday practices. Managing the metabolic conflicts in such flows through further modern interventions became a central concern for the state in following years.

MANAGING METABOLIC CONFLICTS

Increased aquaculture production came into conflict with other resource demands on the lake, most notably as a source of domestic water. Water development had been the key impetus for lake development since 1966, with the initial goal of supplying Manila with water by 1990 (Florendo, 1969). This goal became more urgent with the rapid growth of the Metro Manila region and the emergence of an export-oriented industrial strategy in the Laguna Lake area called the Calabarzon Project (Sly, 1993). Metro Manila primarily depends on the Angat Dam to its north for its water, but during dry El Niño years the city is faced with recurring water shortages that prompt revisiting proposals for alternative water sources.

After decades of studies and plans, the first significant attempt by the LLDA to use Laguna Lake as a source of potable water was made in 1988. This plan considered upgrading lake quality classification from Class C (designated for fisheries) to Class A (for domestic water supply) through a reorganization of lake usage and governance, including stricter standards for effluents flowing into the lake, the phasing out of aquaculture, and the revival of the NHCS's salinity control function (Laguna Lake Development Authority, 1995b). Industrial, agricultural, and domestic pollution and conflicts with the needs of aquaculture and fisheries meant that the plan would require innovations in environmental governance, such as a market-based environmental user fee system to regulate effluents from polluting establishments (Oledan, 2001). From centralized authoritarian models of environmental governance up through the 1980s, Laguna Lake resource management increasingly turned to neoliberal modes in the 1990s.

Despite small, medium, and large industries around the lake numbering more than two hundred thousand in 2005, these new governance mechanisms introduced in the 1990s were able to reduce industrial waste output by half (Natividad, 2005). Yet the issue of regulating household domestic waste presented a more difficult problem amid Manila's horizontal urban expansion and intensified domestic effluent production, which comprised 70 percent of total pollution load in the lake. Informal residents who have settled or have been resettled along the peri-urban northern shores of the lake have frequently been blamed for these increased waste loads, and their eviction had been proposed as a solution to arrest the lake's eventual decline (Natividad, 2005). The state would deploy a similar discourse about informal settlements as primarily responsible for flooding after the 2009 floods and institute an even larger-scale eviction program (see chapter 6).

As part of market-oriented neoliberal reforms in the mid-1990s, water service delivery, among other urban services, was privatized in Metro Manila (Ortega, 2016). The two private water concessionaires that have operated in Metro Manila since 1997 have initiated large-scale efforts to extract domestic water from the lake for urban consumption, partly in response to recurring threats of water shortages during dry years (Esplanada, 2012; *Manila Standard Today*, 2012; Olchondra, 2010). Both concessionaires planned to increase water extraction from Laguna Lake fivefold from the 100 million liters per day (MLD) capacity of the Putatan Plant, operational since 2008 and serving more than a million consumers in southern Metro Manila (Manila Water, 2011; Maynilad, 2010; Olchondra, 2010; Rivera, 2012). The reverse osmosis technology used by the plant, however, needs a certain salinity content in order for water treatment to function, which requires a decrease in saltwater level of the lake through the reopening of hydraulic control and managing pollution levels (Santos-Borja, 1994; Tabios & David, 2004).

With severe water crisis forecast for Metro Manila due to increased demand and constrained supply, Laguna Lake continues to be targeted as a potential source for more water.[8] While early estimates suggested that up to 600 MLD can be extracted from the lake without compromising its multiuse character, the stormwater flow channeled to the lake temporarily from the upstream to avert flooding in the city may be tapped to provide as much as 1,900 MLD (Tabios, 2019). But such plans for domestic water extraction run into conflict with the lake fisheries and are often opposed by fishers and aquaculture producers, who fear the lake's productivity would be severely compromised by reduced salinity.

Recent lake governance has therefore been oriented toward managing these multiple metabolic contradictions between lake and city and attempting to balance its contemporary function as a banal resource frontier with its nostalgic, pastoral image an as aesthetic inspiration. During the LLDA's thirty-second anniversary address in 2001, President Gloria Macapagal-Arroyo expressed this conflict succinctly and unironically: "We want this lake to continue to be a vital resource and instrument for our social and economic growth while serving as inspiration for poetry and romance" (*BusinessWorld*, 2001).

CONCLUSION

Laguna Lake conveniently emerged as a resource frontier at the confluence of the rise of a modern developmental state and the growing pressures of urban growth. The state framed the lake, as a frontier, as simultaneously overexploited and underutilized to institute a broad range of developmental interventions that would render it extractable and exploitable. It served as a laboratory for projects in spatial ordering and state rule driven by modern technologies and practices of unmapping existing socioecological relations. Lake governance during the early years of the LLDA departed radically from previous state attempts to manage the lake's resources, with the prewar concern for fisheries management (Aldaba, 1931a, 1931c; Mane & Villaluz, 1939; Villadolid, 1933, 1934) being replaced by a broader, holistic development agenda that sought to make use of the lake as a source of fish *and* water through modern interventions.

State managerialism permeated the new spaces and bodies of governance, which intersected with faith in scientific and engineering interventions that enabled necessary simplifications of lake socioecologies through experimental technologies and infrastructure on various scales. Knowing, rationalizing, and controlling various components of the lake as a resource became a sectoral concern tied to understanding of particular resource flows. These interventions presented an illusion of control over lake processes that met with various contradictions as complexities continued to elude complete attempts at control, and as one resource use, abstracted from the whole, created fundamental incompatibilities with others. As a frontier, the lake became a site of social and ecological experimentation and innovation, where new regimes of resource governance and control were introduced and tested and their incompatibilities resolved.

The opening scenes from the works of Rizal and Bernal suggest a lake that was alive and brimming with life. Early plans for the lake dreamed up a vision of development and progress for the region and the city. Subsequent transformations that colonized the lake landscape instead reproduced counterimaginaries of the lake as Manila's toilet, a dying lake that needed to be brought back to its pristine condition. How such frontier dreams turned to realities of confusion, chaos, and conflict is the focus of the next chapter.

TWO

Enclosing a Commodity Frontier

ON A JULY AFTERNOON IN 1982, Laguna Lake burned. Around five hundred fishers in a fleet of 140 bancas sailed to a large fishpen in the middle of the lake to dismantle one of the hundreds of aquaculture enclosures that had pushed them out of their fishing grounds. Earlier that day, their fisherfolk association had received instructions to demolish it from the Office of the President, a go-signal that they had been patiently anticipating for weeks. The men slashed and chopped the bamboo poles that fenced in the enclosure and set fire to the hut that had housed the fishpen guards and workers who had kept watch over a valuable resource. The atmosphere was tense as both fishers and fishpen workers watched the fire consume the huts while both sides avoided the violent clashes that had become routine in such encounters. The actual work of dismantling an aquaculture enclosure was difficult and labor intensive, and the day ended with only a quarter of the hundred-hectare fishpen having been removed from the lake (Santos-Maranan, 1982).

This particular fishpen was registered under the Biñan-Magdalena Fishing Corporation, an aquaculture producer owned by an entrepreneur based in Binondo in Manila's Chinatown district. Like many other Manila-based fishpens in the lake at that time, its size and location violated LLDA rules, and it was declared illegal. The government had been attempting to take action in response to the persistent clamor of fishers and accusations that officials were coddling fishpen owners. The owner of this fishpen had been able to delay demolition by going to court to secure a restraining order, finally relenting and promising to dismantle it after one last harvest in 1982 (*Bulletin Today*, 1981b). The case became one of the first high-profile fishpen demolitions, even as many other fishpens declared illegal during this time were not demolished or were rebuilt. Despite being a small, isolated event, the fisherfolk fleet that burned a fishpen was a symbolic, collective reclaiming of the lake and their livelihoods that had been taken away from them.

The battle to dismantle fishpens had raged for around five years at that point, pitting fisherfolk against fishpen corporations and their guards and involving interventions by the LLDA, local politicians, the courts, and even the president. Barely a decade after the first experimental farm introduced aquaculture into Laguna Lake, the landscape had been transformed beyond recognition. Enclosures had taken over half the lake's surface area, with bamboo pole fences jutting out as disorderly eyesores. Many structures were constructed haphazardly, crowding out traditional lake users and effectively privatizing what used to be common waters. With fish stocked considered as future money, the large enclosures were heavily guarded against fishers, viewed as potential intruders and poachers, who in turn carefully navigated around these structures, avoiding direct contact with the pens and occasionally fighting back against guards.

While such encounters have subsided, traces of these scenes remain on the landscape as a reminder of how aquaculture expansion facilitated the enclosure of an urban frontier. The lake became a site fraught with contention and dispossession, where state, capital, and livelihoods intersected with the production of valuable fish commodities. Aquaculture's expansion initiated forms of state regulation of capital's contradictions, a territorial restructuring of lake space for further commodity production, and a deeper subjectification of lake lives to the imperatives of capital.

In this chapter I show how Laguna Lake frontier making is a process of commodity frontier expansion rooted in the ecological dynamics of capitalist accumulation through aquaculture. Building on the accounts of early resource frontier-making projects guided by the state's modern developmentalist agendas in chapter 1, this chapter presents historical and contemporary narratives of the multiple encounters between the logics of capital and agrarian ecologies in a frontier. Specifically, I read Laguna Lake's frontier urbanism as a story of enclosures, ecological contradictions, dispossession, territorialization, subjectification, and resistance. The boom and bust cycles of aquaculture production, aquaculture's urban elite capture, and their impacts on agrarian relations may be understood within a particular logic of how capitalism develops and reproduces through nature. State regulation and the everyday livelihoods of lake dwellers, as we see in this chapter, work with, around, or against such logic, fundamentally reworking socioecological relations and agrarian trajectories in the lake. Capital is an essential element in resource frontier making, and in Laguna Lake, one that is intimately tied to the city.

COMMODITY FRONTIERS, AQUACULTURE, AND ENCLOSURES

Frontiers are made for exploitation and accumulation, as spaces of entrepreneurial opportunity but also as contested sites vital to capital and its spatial logic (Cons & Eilenberg, 2019). These spaces are necessary for capitalism's production of ecological regimes that sustain its development historically. As commodity frontiers, they are spatial manifestations of capital's crises-fixing production of regimes that enroll human and nonhuman natures by producing ecological surpluses that keep system-wide production costs down (Moore, 2011). Two moments accompany surplus-producing ecological revolutions: the creation of resource frontiers or capital's spatial expansion to reduce costs of inputs, also known as commodity widening or the conquest of space in early frontiers, and the development of sociotechnical innovations to intensify capitalization, also known as commodity deepening or the conquest of time in mature frontiers. In turn, these processes produce new natures by joining together appropriation and exploitation and by changing the fundamental organization of socioecological relations (Moore, 2015). Capitalist history may be understood as enrolling cheap labor, food, energy, and inputs at a minimal cost through the extraction of ecological surpluses from widening and deepening of frontiers.

Resource frontier production through commodity widening seeks to produce more commodities with less circulating capital or raw materials and more productive labor. Beyond just increasing capitalization and commodification by extending commodity relations to spaces and bodies not yet commodified, commodity widening is also accomplished by appropriating the "free gifts of nature," including human labor and biophysical nature, that are autonomously reproduced beyond the circuit of capital. In time, these surpluses and the organization of production relations are exhausted through overcapitalization and exploitation inherent in capitalism (O'Connor, 1998; Weis, 2010; Moore, 2015), forcing capital to search for new frontiers or to deepen commodification in existing frontiers through sociotechnical innovations.

Capitalist aquaculture as frontier making is illustrative of the rise and fall of commodity frontiers.[1] As capture fisheries' frontier, aquaculture presents an opportunity to address crises in industrial overexploitation in capture fisheries (Clausen & Clark, 2005; Longo et al., 2015; Mansfield, 2011) by providing new spaces for and new techniques of producing fish and setting in motion the search for ways to offset exhaustion of surpluses in existing fish frontiers. Capture fishing firms invest in aquaculture to overcome limits of declining frontiers, spread risks in fluctuating wild fish catch, and maintain stability and standards in fish flows (Natale et al., 2013). Industrial capture fishing firms often have a greater stake in opening up new frontiers for aquaculture introduction, and they develop infrastructural and knowledge advantages in these new sites of aquaculture development. This has been the case for Manila-based fishing corporations that have expanded their frontiers not only further in the marine spaces of the oceans but also through horizontal integration and investments in aquaculture production to spread risks and enable greater control of fish production.

The expansion of capture fishing firms from Manila involved taking advantage of the opening up of Laguna Lake as a resource frontier (see chapter 1), while simultaneously engaging in sociotechnical innovations to intensify capitalization and produce more commodities in less time (see chapter 3). These processes, however, have inherent contradictions in their attempts to work with particular ecologies of production in frontiers (see chapter 5) and contribute to the eventual exhaustion of the frontier when it is not able to deliver more work. In the lake, as this chapter shows, bursts of very high productivity were followed by near collapses in production, illustrating the mutability and spatiotemporal dynamism of frontiers under capitalism.

In Laguna Lake's history of frontier making, the figure of the enclosure captures the spatial form of nature under capitalism and the state's modern ideology. Aquaculture structures as a mode of enclosure, ranging in size from less than half to more than a hundred hectares, stretched across the lake and enclosed space traditionally accessible to fisherfolk. The enclosures undermined existing sociality built around common rights by restricting physical access to the resource but also established a novel lake economy centered around commodity relations.

Enclosures reconstitute space through the deepening of the commodity economy and capitalist relations and extension of state power through territorialization. Three activities that often accompany enclosures—privatization, dispossession, and subjectification—have spatial characters (Hodkinson, 2012; Sevilla-Buitrago, 2015). Privatization transfers rights to use the lake through delineating and fencing off space, enabling dispossession and loss of access to commonable resources and subjecting people to the commodifying capitalist logic. Enclosures erode people's common rights not only through conscious exercise of exclusion but also through degradation of the commons as a by-product of the accumulation process (De Angelis, 2007) within and beyond the lake.

Enclosure as capital's spatiality encounters a contradiction in its ongoing strategic attempt to extend capital accumulation to further reaches of the world, creating more frontiers but encountering people and society resisting its commodifying logic and destruction of the commons (De Angelis, 2007), as exemplified by the chapter's opening vignette. As in the establishment of many enclosures, the modern state needed to homogenize diverse and complex spaces, rendering them legible, governable, and commensurable through practices of mapping, measuring, regulating space, and keeping boundaries (De Angelis, 2007; Peluso & Lund, 2011; Sevilla-Buitrago, 2015). This territorialization or the use of bounded spaces for control is co-constitutive of the frontier, creating new subjects and relations after capitalist expansion and the establishment of the enclosure (Peluso & Lund, 2011; Rasmussen & Lund, 2018).

Militarization as a form of violent land control shaped resource frontiers and maintenance of enclosures and territory in much of the late twentieth century, when corporate and military power joined forces in the search for and making use of resources (Peluso & Lund, 2011; Tsing, 2005). In Philippine post–World War II history and especially during Marcos's martial law period, these strategies of enclosure involved violence to secure elite control of land that displaced and evicted peasants and rural people (Dressler & Guieb, 2015; Franco & Borras, 2007). It is in this troubled historical context that Laguna Lake's commodity frontier making should also be situated.

URBAN CAPITAL AND THE EXPANSION OF ENCLOSURES

Laguna Lake's landscape of geometric fencing of water is how we would expect an enclosure to look. But the historical-geographical context of its creation and its contradiction deserve further elaboration. In many ways, the lake's landscape represents traditional enclosures of commonable land but is novel in its urban character and the materiality of enclosing fluid environments and livelihoods.

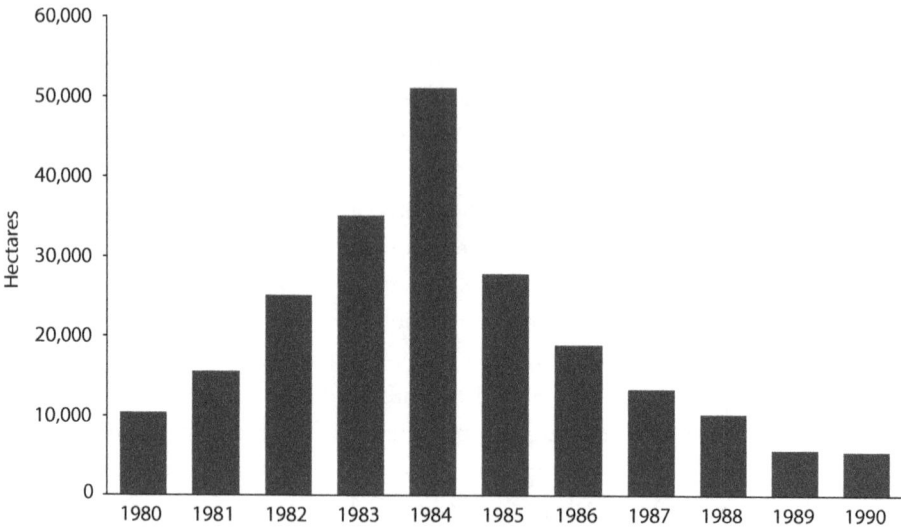

FIGURE 4. Laguna Lake fishpen area, 1980–1990. *Sources:* Delmendo and Rabanal (1982); Eleazar (1992); Laguna Lake Development Authority (1995b).

The success of the first experimental farm spurred the arrival of commercial-scale aquaculture producers at the lake. From 38 hectares in 1971, total fishpen area increased to 4,800 in 1973, 10,400 in 1980, and 35,000 in 1983 (see figure 4), with average fishpen size expanding from 5 ha per operator in 1973 to 290 ha in 1983 (Dela Cruz, 1982; Eleazar, 1992). This trend of a small number of large-scale fishpen operations occupying hundreds of hectares of lake area, which began in the late 1970s, continues up to this day. The intrusion of urban capital shaped the trajectories of aquaculture in Laguna Lake over the next half century.

Alongside the demonstration of fishpen feasibility in the lake, the LLDA planned to identify qualified fisherfolk for fishpen development. Local government units were to recommend to the LLDA those constituents to be granted permits for construction. In the absence of clear implementing rules until 1976, however, investment interest swelled from well-capitalized individuals based in Manila and the province of Bulacan to its north, who took advantage of affordable leases and the ease of acquiring permits (Delmendo & Rabanal, 1982; Ruaya, 1994). Confusion over whether the LLDA or the municipal government had the authority to issue permits led to a situation in which both were granting leases separately, allowing further entry of fishpen operators with minimal regulation.[2] Meanwhile, aquaculture remained inaccessible to fisherfolk due to the high construction and operating costs of fishpens (Dela Cruz, 1982; Delmendo & Gedney, 1976; see table 2 for a contemporary comparison between capital-intensive fishpen production and small-scale cage production more accessible to fisherfolk).

With good water conditions conducive to biannual fish harvests and without devastating typhoons, the fishpen rush intensified between 1977 to 1984, peaking at a record area of 51,000 ha or more than half the lake's area. "I have recovered my investment with profit after the first harvest," claimed a producer in 1982, suggesting that fishpens were highly profitable during these early years (Cruz, 1982, p. 4). A wide variety of Manila-based investors established fishpens in the lake, including politicians, military officials, film celebrities, professional basketball players, foreigners, and others who had enough capital to start commercial operations (Cruz, 1982; Ofreneo, 1980; Santos-Borja & Nepomuceno, 2006). By 1981, it was commonly reported that "former senators, assemblymen, military men and others who had right connections" owned many of the largest fishpens in the lake (Manipol, 1981a, p. 5).

The earliest investors and largest operators were individuals and fishing corporations based in the northern Manila Bay coastal towns who had well-established deep-sea or fishpond aquaculture investments and know-how in fisheries (Jose, 1994a). The development of capitalist aquaculture in Laguna Lake should be situated within the responses of these corporations to historical change in industrial capture fisheries in the Philippines, a sector beset by stagnant production, rising fuel costs, and a constant search for new fishing frontiers (Green et al., 2003). Fishing firms based in Manila's wholesale fish market, through which close to half a million metric tons of fish pass annually, are central to Philippine industrial fisheries. These firms, with origins in the adjacent northern Metro Manila fishing centers of Navotas and Malabon, exert oligopolistic control over various stages of the fish commodity flows to Metro Manila, from production (industrial deep-sea fisheries and large-scale lake aquaculture) to exchange (fish market brokers).[3] In 1992, the five largest fishing firms accounted for 70 percent of the capture fisheries' haul in the fish market (Tiambeng, 1992), and two-thirds of the fish were handled by the eight largest brokers, which these firms also owned (see chapter 4).

These firms expanded from their origins, a mixture of the old elites and the new rich of Malabon and Navotas, to take advantage of changing fish economies. Both were pioneers in fishpond aquaculture production, which was traditionally a rural elite venture (Villaluz, 1950). In Malabon, the *naturales* (native elites) and the *principalia* political class favored by the Spanish

TABLE 2 Comparison of Fishpen and Fishcage Production

Category	Fishpen Production	Fishcage Production
Average size of production unit (2010)[a]	27.9 ha	0.6 ha
Total area occupied (2010)[a]	11,430 ha	2,000 ha
Total number of operators (2010)	410	2,920
Ownership	Mostly absentee urban entrepreneurs registered as individuals or corporations; some large fishing companies and a few fisherfolk cooperatives	Mostly lake residents; some enter into partnerships with urban financiers
Pen/cage investment requirements (2006)[b]	P3.72 million for a 50 ha pen P0.74 million for a 5 ha pen	P0.16 million for a 1 ha cage
Total costs (2006)[b]	P7.46 million for a 50 ha pen P1.10 million for a 5 ha pen	P0.12 million for a 1 ha cage
Net revenue (2006)[b]	P6.54 million for a 50 ha pen P0.43 million for a 5 ha pen	P0.19 million for a 1 ha cage
Fixed assets (poles, nets, caretaker huts, etc., 2006)[b]	P3,719,500 for a 50 ha pen	P195,090 for a 1 ha cage
Fixed assets as percentage of total costs	49.8% for a 50 ha pen	129.5% for a 1 ha cage
Production volumes of species grown (2010)[c]	Total: 48,500 MT (milkfish, 43%; bighead carp, 31%; tilapia, 25%)	Total: 12,300 MT (tilapia, 86%; bighead carp, 14%; milkfish, 0%)
Stocking density (2007)[d]	35,000 pieces/ha	133,000 pieces/ha
Frequency of harvest in a year (2007)[d]	16	1
Yield (production/area in 2010)	4.24 MT/ha	6.15 MT/ha
Labor arrangements (permanent waged and seasonal hired labor)	Permanent: caretaker and wage laborers (average of 14 for a 50 ha pen); usually migrant labor Seasonal: seiners for preharvesting	Permanent: 0–1 wage laborer for a 1-ha cage; usually village labor Seasonal: village labor hired for many tasks

[a] From LLDA database.
[b] From Israel et al. (2008)}; 1 US dollar is equivalent to 48 Philippine pesos at 2006 exchange rate
[c] From Philippine Statistics Authority OpenSTAT database.
[d] From Tan et al. (2010).

colonial government shifted from tobacco and sugar refining in the nineteenth century to milkfish fishpond aquaculture at the turn of the twentieth century, after taking over lands from friars during American colonial rule and dominating urban-oriented fishpond aquaculture production for decades to follow (Magno, 1993). Taking advantage of opportunities opened up by Laguna Lake aquaculture in the 1970s and armed with prior milkfish aquaculture experience and knowledge, several of the fishpond owners extended their investments to pen aquaculture in Laguna Lake as the first investors who benefited from the initial resource rush.[4]

Navotas families invested in industrial fishing ventures, taking over from the Japanese, who controlled Philippine commercial fisheries in the first half of the twentieth century (Morgan & Staples 2006; Ofreneo, 1980). Through technological transfer from and joint ventures with Japanese firms, Navotas fishing firms expanded their operations between the 1950s and 1970s (Ofreneo, 1980; Spoehr, 1984). One of the largest fishing corporations in the Philippines started out with one Japanese deep-sea trawler in 1963 but has since diversified into freshwater aquaculture, tuna fishing off Papua New Guinea (Havice & Reed, 2012), food processing, and real estate. Another large fishing corporation started out as a small family enterprise with one vintage fishing boat in the 1970s before expanding its trawl fishing operations and recently diversifying to retail. These deep-sea fishing firms continue to operate the largest pens in Laguna Lake, investing in lake aquaculture to supply the urban fish market with freshwater fish to complement fluctuating marine fish volumes of urban staples, especially during the lean months of October to February.

Prior to the introduction of aquaculture, Laguna Lake fisheries were primarily village based and small scale. The production of Laguna Lake as a commodity frontier, however, opened up opportunities for fishing corporations and entrepreneurs to spread the risks and limitations of deep-sea fishing or fishpond culture while diversifying their sources of fish supply. This fits with observations about aquaculture-capture fisheries interactions, which show how horizontal integration balances production fluctuations, provides diverse products, and addresses limits in one sector (Natale et al., 2013). Favorable political connections and weak regulation characteristic of early frontiers, combined with Laguna Lake's unique ecological characteristics, allowed these early-entry firms to appropriate the high ecological surplus of a plankton-rich production environment.

The fishpen boom brought reports that conjured images of frontier chaos in the lake: proliferation of registered and unregistered fishpens, which occupied navigation lanes, ignored standards for stocking densities, and overfed their fishpens with redundant supplemental feeds (Rivera, 1987). Yet the highly productive conditions would not last long (see figure 2 in chapter 1). Prior to 1976, a twice- or thrice-yearly cropping cycle was common practice. This proved more difficult in the 1980s, when culture that once took three months extended to as much as eight to ten months (Palisoc, 1988; Richter, 2001). Studies proposed a variety of explanations to account for how aquaculture proliferation in the lake had caused its own decline in productivity. Pen structures affected water circulation because the drag effects on water flow by the fences led to a reduction in the dissolved oxygen vital for fish life (Delos Reyes, 1993; Lasco & Espaldon, 2005). Fish excreta and the supplemental feeds also contributed to increased ammonia and nitrogen loading in the lake, which contributed to algae blooms and fish mortality (Delos Reyes, 1993; Richter, 2001; Zafaralla et al., 2005), while detritus from fish and fishpen structures increased in comparison to phytoplankton, which reduced fish intake of lake food (Richter, 2001). Very high stocking rates of up to a million fingerlings per hectare also contributed to fish morbidity and mortality and thus low productivity (Laguna Lake Development Authority, 1999; Richter, 2001).

Total fishpen area in the lake began to decline after the 1984 peak, falling to a tenth of peak size in 1990 (see figure 7 in chapter 3). Two sets of reasons explain this decline. The first pertains to limnological changes brought about by fishpen expansion, the hydraulic control of saline backflow, damage by typhoons, and other activities that caused a decline in primary productivity or increase in algae blooms, which made fishpen production less profitable. Decreasing fish yields in aquaculture and capture fisheries are tied to the complex interaction of factors that determine primary productivity. In the case of a eutrophic Laguna Lake, the presence of nutrients is important in producing the phytoplankton that the fish consume for growth. Yet too many nutrients can also cause algae blooms, which contribute to large-scale fish kills during algae die off.

The second set involves state efforts at regulation, most notably the dismantling of fishpens and the establishment of a zoning plan. Both illustrate the dynamics of materiality of nature, modernity, and the contradictions of lake production as an exhausted commodity frontier. The lake failed to sustain the delivery of high ecological surpluses. This required further state intervention to attempt to restore its productivity and profitability by regulating the socioecological conflicts and contradictions that resource frontier production created.

RESOURCE RUSH AND THE FISHPEN CONTROVERSY

The introduction of aquaculture was accompanied by institutional changes in lake space rights, usage, and enforcement. Before the 1970s, fishers could set up lines or nets almost anytime and anywhere in the lake based on informal rules rooted in mutual respect for rights to fish (Eleazar, 1992). There have been government regulations pertaining to limits of mesh sizes, closed season for drag seines, use of destructive gear, and fisherfolk registration through the municipal government (Eleazar, 1992; Villadolid, 1933, 1934). However, territorial exclusion by limiting access to particular sections of the lake was uncommon, partly due to the fugitive character of fish, which requires mobility of fishing boats and a variety of gear operating at different territorial scopes.[5]

Fishpen development highlighted the novel concept of an exclusive and bounded space for production in the lake. Fisherfolk made use of fish corrals, structures that enclose fish somewhat similar to fishpens, throughout much of the twentieth century but without having to pay rents or fees. Fishpens were different because they occupied vast tracts of lake space, usually in the most productive portions of the lake, and their ownership had been formalized by their LLDA leases.

Since fishpens are costly investments that produce highly profitable milkfish, many operators guard their structures vigilantly. In the 1970s and 1980s, hiring armed guards was a common practice to ward off poaching of fish and sabotage of nets by fishers displaced from their fishing grounds (Cruz, 1982; Delmendo & Gedney, 1976). Reports suggested that these armed guards were either military personnel working for military fishpen owners or hired through a security agency (*Bulletin Today*, 1984c). Fishpens also made navigation physically difficult and costly for fishers, who now had to maneuver around the enclosures to reach a much-reduced fishing ground. Fishers recall guards shooting at them if their boats or the nets that they set drifted close to the fishpen perimeter.

At the height of the fishpen sprawl, Cruz (1982) and Santos-Maranan (1982) documented incidents of guards beating up fishers who strayed too close, pointing guns at their heads, and extorting money or fish before allowing them passage in the waterways. In one case, after an encounter with masked guards that left him bleeding from being hit in the face with a rifle, one fisher was threatened that something worse would happen if the incident was reported (Santos-Maranan, 1982). But fisherfolk rarely had the opportunity or the desire to report such incidents individually due to fear of retaliation and distrust of authorities.

Fishers lost access to lake space that was once open to them and had been enclosed by fishpens. Fisherfolk took action by forming alliances demanding action from the government. The Laguna-based Samahang Mandaragat ng Lawa ng Laguna, Ink., for example, met with then minister of defense Juan Ponce Enrile and aired their grievances against the fishpen encroachment and the harassment and attacks (Santos-Maranan, 1982). Fishers also took matters into their own hands by demolishing illegal fishpens, having discovered the futility of demanding action from government officials, whom they knew operated some of the largest fishpens in the lake. Fisherfolk participation in fishpen dismantling would continue after the 1982 landmark demolition that opened this chapter, with a dash of pessimism that the state would demolish pens on the one hand but register new ones on the other (Jose, 1994a).

Accounts from lakeshore villages in the 1980s suggest simmering unrest was ready to break the surface. Villagers were organizing vigilante groups and taking action by arming themselves and forcibly dismantling fishpens at gunpoint. A fisher wrote to a national newspaper in 1983 expressing concern over the growing turmoil in the village: "The people in our *barangay* are now becoming restless, impatient with the way the government is handling the situation. They are now massing anew and planning to attack existing fishpens and these new ones being constructed in our area. We do not want these things to happen. We do not want bloodshed. It could be that this is what it would end. Perhaps the worst is yet to come" (Bigornia, 1983b, p. 4).

Bloodshed did occur in early 1984 when two fisherfolk from Binangonan were shot and killed by armed fishpen guards (*Bulletin Today*, 1984a). The LLDA responded by requesting assistance from the armed forces to help investigate and conduct nightly patrols in the area. In a display of populist response, President Marcos then ordered the police "to protect the small fishermen against 'big fellows' blocking the dismantling of illegal fishpens" (*Bulletin Today*, 1984b, p. 1). The increased presence of the police and military in the lake was not accidental, coinciding with state efforts to quell communist insurgency fueled by unrest surrounding the fishpen controversy. It is not surprising therefore that authorities had often called fisherfolk who fought back "subversives," a term associated with state enemies under martial law (Santos-Maranan, 1982).

The contrast between the fixity of aquaculture and the mobility of capture fisheries stoked the tension between fishers and fishpens. The former viewed the latter as outsiders that displaced them, the legitimate users of the lake. "No one owned the lake before. Only the Laguna Lake people used the lake," commented Max, a boat owner in his fifties, recalling a time prior to the peak fishpen rush.[6] "This lake used to be for the poor only," said Gardo, a gillnet fisher.[7] He and other fishers recall a time prior to aquaculture and state intervention in fish resources when the lake was "unobstructed" and they were "free to move and make a living." Despite regaining some of the fishing grounds lost to fishpens after the peak sprawl, fisherfolk continue to see their long-standing effects in the curtailment of this freedom to move around and make a living.

Fisherfolk identities in this context are mobilized to stake a claim to the lake against aquaculture expansion and dispossession. The reference to fisherfolk as the "legitimate" and "original" users of the lake emphasizes that their claims to the lake are a product of their rootedness in place, encompassing all kinds of people making a living from the lake. These claims, which respect the freedom to move and make a living while unifying a diverse range of users, are contrasted with outsiders who extract from the lake for others' benefit.

Violence shapes access and is a crucial component of establishing enclosures and marking territory that produces new environmental subjectivities (Peluso & Lund, 2011). It permeates frontier spaces, where property regimes are often unstable and authority is contested (Rasmussen & Lund, 2018). The experience of violence or its memory in Laguna Lake enclosures during the peak resource rush is perhaps best described by Santos-Maranan's (1982) narration of an encounter with fishpen guards by Tibo, who was taken into the hut after accidentally sailing near the structures. With guns aimed at him, the guards interrogated him, only releasing him after finding out he personally knew the owner of the fishpen. He was allowed to go back to shore but not before the guards confiscated some of his catch for their own consumption. Fisherfolk like Tibo had to constantly navigate the threat and potential of violence that represented the fishpen structures in the lake.

It was inevitable that fishing boats would encounter pens, especially since fugitive fish apart from those stocked by operators are attracted to fishpen enclosures because of the calm and sheltered waters within and immediately outside these structures (Tan et al., 2010). Pens even become the de facto owners of nonstocked fish that grow and are harvested inside their pens. A fisher observed this pattern: "Fish congregate near their nets. That is why we can only catch fish if we stray near the pens. But they keep on driving us away."[8]

Fisherfolk are often seen to infringe on the territorial bounds of pens. Given the fluid circulation of water, gear and nets set near pens sometimes crash into or damage pens (and vice versa). This has built a sense of mistrust between fishpen caretakers and guards and fishing crew. In response, fishpen guards exercise their power to exclude by demanding remuneration whenever there is damage to their nets, levying taxes or fees on passing fishing boats and generally viewing all passing fishers with suspicion. The use of fireworks and firearms by fishpen guards to drive fishers away has resulted in tension and recurring violent encounters.

Over time, fishers came to view pen-fisher antagonistic relations as part of their everyday encounters in the process of making a living. Fisherfolk distinguish between nice pen operators or caretakers, who do not mind if they fish nearby, and those who are less tolerant. They describe fishpen attitudes that range from just strict (*istrikto*) and ill-tempered (*masungit*) to bad or nasty (*salbahe*). Fishers, for example, know if a pen owner will allow them to drift close by the pen laborers' hand gestures or their use of guns or fireworks:

> There are large fishpen operators that are kind. There are nasty ones, like this operator from Malabon who has killed many. So, few fishing boats drift close to that pen. There are nice ones though who let us in their pens to clean them from knifefish invasives before they stock again.[9]

The mistrust between the two groups is complicated by the intentional violation of the fishpen's property rights by some fishers who see the poaching of pen fish as retribution for being "too strict," aside from its being a means to arrest hunger and provide income for their daily needs. Poaching by "seamen" or "pirates" has become a fisherfolk strategy to gain access to fish that became the property of pens. The invisibility and stationary character of fish within enclosures have made poaching easy, common, and lucrative for fisherfolk.

REGULATING UNREST AND THE TERRITORIAL RESTRUCTURING OF THE LAKE

The state—through the LLDA, local governments, and the national government—attempted to rein in frontier chaos and lawlessness in Laguna

Lake, increasingly framed within broader fears of a rebel insurgency infiltrating lake unrest.[10] With the unruly expansion of fishpen structures that, according to the president, looked "like a slum" (Ng, 1983), the centralized authoritarian state responded with a host of strategies to address aquaculture's unintentional transformation of lake socioecologies: demolition and zoning ("rationalization") and building fisherfolk access to fisheries and aquaculture ("democratization"). The first significant regulatory mechanism that addressed pen conflicts involved national government intervention through the centralized power of the Office of the President.

Marcos issued several letters of instruction, including in 1978 (LOI 769) and 1982 (LOI 1239), that tasked the LLDA and other state agencies to oversee the demolition of illegal fishpens (Laguna Lake Development Authority, 1978a, 1982, 1983). Marcos deployed task force teams led by the police and military with the goal of "cleaning up" the lake and freeing it from obstructions (*Bulletin Today*, 1984d). But as news and LLDA reports attest, dismantling the fishpens proved to be much more difficult than anticipated. In 1981, after a renewed order from the LLDA general manager, it was reported that not a single illegal fishpen had been demolished (Manipol, 1981b), and by 1983 fishpens were still expanding operations (*Bulletin Today*, 1983a). The fisherfolk fleet that demolished and burned the fishpen described in the chapter's opening was a landmark yet incomplete and isolated event.

Groups of fishpen operators also went to court to file restraining orders, which temporarily suspended demolitions (*Bulletin Today*, 1981a, 1982a). However, it was the strong political clout and "right connections" with state and LLDA officials—some operators were state officials and military men who had connections with Marcos and subsequent presidents—that frustrated systematic dismantling of fishpens (Jose, 1994a; Tribdino, 1996). Despite Marcos's strongman rhetoric and authoritarian interventions, including deploying the police and military in the lake, fishpen expansion continued unabated until the ecologically driven bust in aquaculture production of the mid-1980s.

The second regulatory intervention, territorial regulation of the lake through zoning, emerged in tandem with the demolitions. In 1983, the LLDA proposed the creation of the Fishery Zoning and Management Plan (ZOMAP) as a longer-term solution to the pen sprawl problem, with implementation initially set for the following year (Laguna Lake Development Authority, 1983b). Based on the estimated carrying capacity of the lake, fishpen area was set to be reduced to 21,000 hectares and located in the offshore

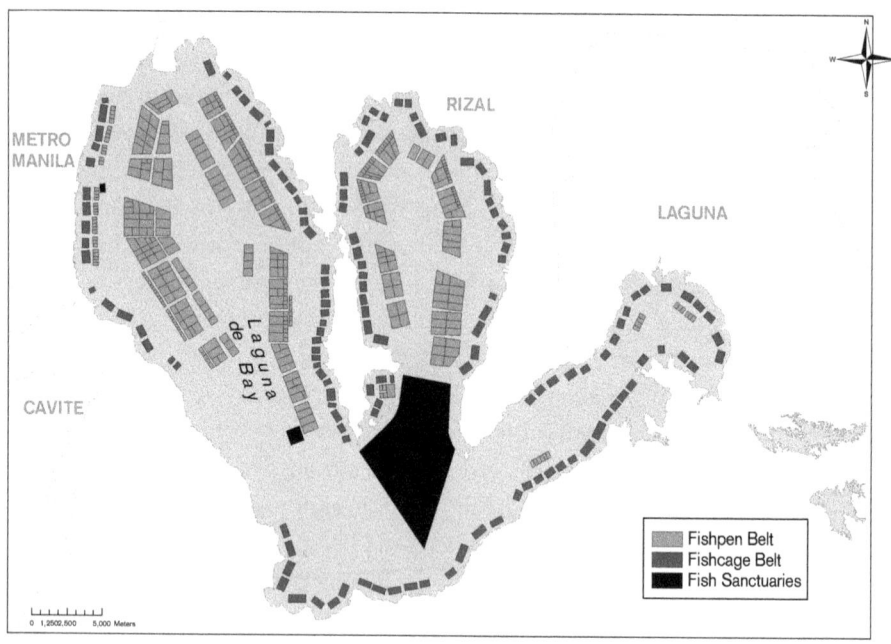

MAP 3. Revised ZOMAP, 1999. Map by Patricia Anne Delmendo.

sections. The ZOMAP also delineated fishcage belts closer to the shore, navigational lanes, and fishing grounds with the goal of apportioning the remaining 70,000 ha of space to displaced lake users (see map 3).

By delineating specific lake space for use by fisherfolk, cage owners, and fishpen operators, the ZOMAP centralized the lake as a governable resource with distinct spatial zones represented by geometric belts. Through aquaculture, the state made fish production in the lake more legible and easier to manage, count, and survey than the fugitive character and mobile institutions that characterized traditional capture fisheries. With the failure of initial aquaculture intervention, this regulation of aquaculture through zoning (and rezoning) can be seen as a form of (re)territorialization of state power (Peluso & Lund, 2011; Vandergeest et al., 1999; Vandergeest & Peluso, 1995).

The ZOMAP as a novel intervention had a threefold aim that fits squarely within the modern production of the lake: to regulate fishpen expansion and democratize fisherfolk use by increasing area for fisheries while further improving fish productivity. It sought to provide a technical, territorial solution to the fishpen controversy by reworking the contradictions in aquaculture as a new form of socioecological arrangement when conflicts threatened to reduce production. Amid social unrest, the ZOMAP promised a 60 percent improvement in fish production for a year and a threefold increase in the subsequent three years (*Bulletin Today*, 1983c). It effectively demonstrated high modern state planning in a commodity frontier: creating order and improvement through the modern language of rationalization and democratization (Scott, 1998). The ZOMAP intervention describes a form of institutional restructuring of the lake to solve the fishpen crisis generated by the contradictions of capitalist aquaculture's commodity production (Bridge, 2000).

The ZOMAP had a depoliticizing effect, despite its very political framing. Rather than challenging the power of fishpens, the ZOMAP territorialized the contentious presence of large-scale, capitalist aquaculture in the lake, effectively legitimizing fishpens as long as they met zoning guidelines. A fisherfolk group expressed apprehension that "because illegal fishpen operators are moneyed and influential, they'd eventually legalize their position" (*Bulletin Today*, 1982b, p. 32). As succeeding decades would show, fishpen owners were able to not only resist demolition but also get away with not paying leases and with operating beyond the 50-hectare limit set by LLDA regulations through the creation of dummy or shell corporations (Yap, 1999). It has also been reported that they have influenced through bribes the classification of which zones are declared legal and illegal on the map(Jose, 1994b).

The ZOMAP rendered technical the fundamentally political issue of access to the lake (Li, 2007). Through rationalization of lake space and democratization of ownership (Laguna Lake Development Authority, 1999), the zoning presented the solution to a political problem as a matter of spatial ordering of contentious access regimes through belts and zones. It sidestepped fundamental issues of access while providing a form of state intervention that accepted that fisherfolk needed to coexist with pens. Conflicts between fishers and pen operators could be solved by spatially ordering production and putting their livelihoods in their proper places. This would lead to improved fish production, presenting a win-win solution for all parties concerned, without preventing fishpens continuing to earn from the lake (Laguna Lake Development Authority, 1995b; Samonte, 1983).

Legitimizing fishpens in the lake had also been framed through their economic contributions (Samonte, 1983) and role in supplying fish for the urban population (Bigornia, 1983a), arguments that are still made when matters of fishpen demolition arise. In 1981, an assemblyman who owned fishpens in the lake reiterated that fishpens contributed to lower fish prices in the city to oppose plans of dismantling his structures, while highlighting that fisherfolk also benefited indirectly from aquaculture when fish escaped pens during typhoons (*Bulletin Today*, 1981c). Marcos would also repeat the statement that fishpens supplied 80 percent of the fish in Metro Manila and nearby regions (Ng, 1983). As part of his authoritarian developmentalist model, the necessity for compromise and efficient planning in the lake was a reaction to increasing urban pressures and various resource demands on the lake (Marcos, 1983). The urban economic connection justified downplaying the more radical promise of bringing the lake back to the people marginalized by fishpen aquaculture.

The third state strategy of regulation was promoting fisherfolk access to aquaculture through democratization. This entailed redistributing space once occupied by dismantled fishpens, turning them over to fishers and "thereby democratizing the bounties of the lake" as promised (*Bulletin Today*, 1983b). It also involved organizing fishers into cooperatives to enable them to manage fishpens on their own. Marcos brought his national cooperative program, the Kilusang Kabuhayan at Kaunlaran or KKK (Movement for Livelihood and Progress), to the lake (Cardenas, 1983; Ng, 1983), installing former Farm Systems Development Corporation head Teodoro Rey as LLDA general manager. Marcos issued LOI 1325 in 1983, which organized small fisherfolk into associations called the Kapisanang Kabuhayan sa Kalawaan (Lakewide Livelihood Association) as part of the Laguna Lake Cooperative Development Program, funded by external loans (*Bulletin Today*, 1984d; Laguna Lake Development Authority, 1982). It aimed to develop "a consciousness for shared responsibility and group action" (*Bulletin Today*, 1984e, p. 1) among fisherfolk to align with his vision of self-reliant, entrepreneurial livelihoods in the New Society (BFAR, 1981).

A similar attempt to make aquaculture accessible to fisherfolk had been made years earlier through the Laguna de Bay Fishpen Development Program (LDBFDP). Funded by a $95 million loan from the ADB and Organization of the Petroleum Exporting Countries, the program sought to provide loans for fisherfolk cooperatives to enable them to operate 10 ha fishpens (Laguna Lake Development Authority, 1977, 1978a, 1979, 1980, 1981). By 1985, the LLDA had issued permits to 1,373 members in eighty-eight fisherfolk cooperatives (Laguna Lake Development Authority, 1985).

Despite significant funds being injected into the two projects, both were massive social development failures. The Cooperative Development Program ceased in 1986 with only 105 out of 400 expected cooperatives formed because of internal disunity and uncertain feasibility of the projects (Laguna

Lake Development Authority, 1986). None of the LDBFDP's target goals were met, with only 2 percent of projected outputs delivered by 1988. The cooperative component encountered disastrous results, putting fisherfolk cooperatives in significant debt that reached more than a million pesos per fishpen (Yap, 1999). Destructive typhoons and the fishpen sprawl that occupied the most productive parts of the lake limited the ability of beneficiaries to derive profits from their operations (Eleazar, 1992; Yap, 1999).

Rather than being driven by a monolithic state vision, aquaculture improvement schemes demonstrate that the state is composed of various individual or group interests that have multiple discursive formations and possess different capacities for action. The interest of the LLDA in managing and reducing unruly pen expansion through the ZOMAP, for example, clashed with the interests of local government units and politicians, who sought to derive benefits from granting permits or maintaining ties with pen operators. The LLDA, local government, and national government pursued complex relations with pen operators and viewed their future in varying ways. Territorialization in frontiers emerges not from a preexisting state power but instead is generative of authority (Rasmussen & Lund, 2018). In the case of Laguna Lake, the LLDA's tenuous authority was strengthened through the act of delineating space through zoning and regulation. As a fledgling state body that was transformed from a development-oriented agency into a regulatory body, its power emerged as a result of its ability to collect taxes and fees and to regulate highly profitable resource production.

Despite the intentions of making fishpen aquaculture accessible to the poor, the prohibitive cost of its construction and operation and the institutional confusion between the LLDA and local governments in granting permits led to the displacement of fisherfolk from the lake by urban entrepreneurs and corporations. On the other hand, the state also introduced fishcages as a viable alternative technology to fishpens in response to social unrest. This in turn transformed agrarian relations and subjectivities in villages traditionally dependent on capture fisheries.

AQUACULTURE AND AGRARIAN CHANGE IN LAGUNA LAKE VILLAGES

States continue to promote aquaculture as a development tool for producing more fish, improving livelihoods, ensuring food security, and reducing poverty (Belton & Thilsted, 2014; Bene et al., 2016). Aquaculture transforms agrarian relations in places where they develop, creating varied, if conflicting, political ecologies as accumulation, differentiation, and dispossession intersect on these frontiers.[11] In many cases such as Laguna Lake, however, benefits of the projects end up being captured by wealthy absentee urbanites or local elites instead of their target beneficiaries (Adduci, 2009; Belton & Little, 2011; Toufique & Gregory, 2008), creating a production landscape wherein capitalist aquaculture firms co-opt or coexist with small-scale aquaculture producers in production or in backward and forward linkages (Goss et al., 2001; Ito, 2002; Vandergeest et al., 1999).

In Laguna Lake, the shift from capture fisheries to small-scale aquaculture involved not only radical changes in livelihoods and village relations but also the forging of new economic connections with large-scale capitalist aquaculture. The commodity frontier is thus shaped not only by intrusion of large-scale capitalist configurations introduced by the state into everyday agrarian lives but also by the incremental initiatives (Ghosh & Meer, 2021; Li, 2014) of people adopting aquaculture, inserting themselves in market relations, and building economic connections with capitalist aquaculture. These complex livelihood landscapes on the frontier speak to the agrarian question of how capitalism is transforming class relations and generating differentiation among producers (Akram-Lodhi & Kay, 2010; Bernstein, 1996; de Janvry, 1981; Kautsky, 1988).

The reproduction squeeze (Bernstein, 2010) experienced by fisherfolk as a result of capitalist enclosures and dispossession has brought questions of what happens when rural peoples are "depeasantized" and become surplus populations, often absorbed by the city through cycles of migration (Ghosh & Meer 2021; Li, 2010). The enclosure is thus termed "the midwife of the capitalist city" (Hodkinson, 2012, p. 500), as it initiates expulsion of rural peoples to bring them to the city as proletariats. In many cases this proletarianization is partial or incomplete, resulting in agrarian economies that see a mix of capitalism, petty commodity production and noncapitalist livelihoods, and people engaged in urban informal economies.

In Laguna Lake, aquaculture has brought dispossession and separation of fisherfolk from the lake as commons, creating an army of potentially surplus labor similar to landless peasantry divorced from their means of production. But the same process has also created new economic subjects driven by adoption of small-scale aquaculture, a shift that resembles in many ways family labor–reliant petty commodity production with linkages to capitalist fishpens and occasional wage labor. Small-scale cage aquaculture coexists with traditional capture fisheries livelihoods despite the dispossessing effects of enclosures. All three types of fish production—capitalist fishpen aquaculture, small-scale cage aquaculture, and capture fisheries—occur in Laguna Lake, producing rooted socioeconomic networks surrounding the fish commodity. The cases of two villages at the lake show how this tenuous coexistence has shaped agrarian fates.

Kalinawan, a village in the municipality of Binangonan located on the north-central coast that is considered the lake's fishery center, engaged primarily in capture fisheries as a source of livelihood prior to the introduction of cage aquaculture in 1980.[12] Large fishing boats caught various kinds of indigenous fish and snails through motorized push nets and drag seines, and along with smaller fishing boats that used gillnets, long lines, and cast nets, provided villagers with a means of subsistence. These large motorized push net and drag seine boats, owned by a few wealthier village households, employed a crew of more than a dozen. With the advent of aquaculture and increased enforcement of the 1998 Fisheries Code that banned fine-mesh nets, some of these boats were converted for ancillary pen work, such as seiners for harvest-ready fish in pens and traders for hauling and transporting pen fish to landing ports.

The adoption of small-scale cage aquaculture in Kalinawan, however, required changes different from the institutional arrangements in capture fisheries. A share tenancy system exists in the Laguna Lake fisheries, similar to other marine fishery institutions (Carnaje, 2007; Spoehr, 1984), which parallels the *kasama* system in rice farming in the Philippines (Aguilar, 1989; Ofreneo, 1980; Takahashi & Fegan, 1983). Wealthier villagers own the boat and/or nets, and village-based crews are the share tenants (*kasama*) or crew members (*tauhan*), with members sharing half of the net earnings per fishing trip. Crew labor is recruited based on kinship and patronage ties, which are strong between crew and boat owners and between crew and boat captains. Crew members get credit for daily household expenses from boat owners in exchange for their regular labor on the boats. On larger boats, captains make fishing decisions, with approval from the boat owners, in terms of who to hire, where to fish, when to unload, and so on.

Kalinawan adopted cage aquaculture as a result of the extension efforts of the LLDA and nearby SEAFDEC research station to disseminate tilapia seeds and breeders.[13] Villagers have embraced aquaculture more fully (around 95 percent of households) than neighboring villages for a host of reasons: locational advantages, more concentrated pre-aquaculture wealth distribution, productive water conditions, an initially active cooperative established by the Cooperative Development Program, and personal ties of villagers with research station staff. The rural cooperative served as the initial vehicle of aquaculture technological and knowledge transfer. Referring to himself as the grandfather of tilapia culture in the village, Romy narrated how aquaculture first came to Kalinawan in 1980 in a rather incidental way:

> SEAFDEC offered transfer of technology for tilapia breeding. That was how we started. I was a member of the *Samahang Nayon* (Rural Cooperative) and no one was willing to attend seminars. I volunteered, since I was a member of the Board of Directors. I attended seminars in [nearby] SEAFDEC for a year. Then I taught people here how to breed tilapia.[14]

Tilapia are reproduced through the manipulation of the male-female ratio of sexually mature breeders within hapa net cage enclosures.[15] The transfer of knowledge of tilapia breeding through experience, observation, and sharing was relatively simple and easy for villagers like Romy, who were limited only by their ability to secure enough funds to purchase nets and poles for their cage nurseries (*semilyahan*). Many cage producers were able to invest in *semilyahan* using money borrowed from better-off relatives or saved from urban off-farm work.[16] Kalinawan came to be known as one of the fish fingerlings centers of Laguna Lake, supplying newly hatched tilapia and bighead carp to pens and other cages throughout the lake. Villagers often point to large houses as proof that cage nurseries have brought certain households more prosperity.[17] In an apparent reversal of the depeasantization process, stories of villagers who worked in wage labor in the city only to return to the village for aquaculture were common:

> After we began culturing fish here, our lives changed. You don't need to worry anymore about what to eat or how to put kids into school. Our lives became better when there were fishpens here. You can eat three times a day, construct a house, go wherever you please. If you were only fishing, you will not be able to save and what you eat will not be enough.[18]

> My *kumpare* [coparent through baptism] used to work in Mitsubishi in the city. When he learned how to breed tilapia, his cages multiplied. He got startup money when he resigned from the company. His house is tall. It used to be a small house he inherited from his parents but he made it tall. If he stayed on with work in the mainland, he probably would not be able to build that house.[19]

FIGURE 5. "Visiting" a fishcage aquaculture nursery (*semilyahan*), 2012. Photo by author.

Cage aquaculture in Kalinawan (see figure 5) involved a radical reworking of ownership and production that was distinct from the share tenancy system that prevailed in capture fisheries. For the first time, many producers were able to own their means of production, instead of relying on crew work in fishing operations owned by a few households. They have become their own bosses, who make their own decisions and employ strategies to earn: "I may not have money like my professional friends but I have people working for me. I am the boss, I have a small business."[20]

Cage nursery producers often engage in share-system partnership with urban, middle-class investors, who provide money in exchange for their labor and knowledge, creating a different stream of urban connections.[21] The deconcentration of ownership to several households in the village also increased the predominance of fixed-wage labor for stay-in or all-around work in cages.[22] Village relations shifted from group-based share tenancy in capture fisheries to more individualized production in cages. The patronage relations between a few boat owners and several fishing crews were transformed into relations between several cage producers and wage or hired laborers. A few wealthier households who formerly owned the large capture fisheries boats also engaged in the more capital-intensive and higher-return bighead carp hatchery:

Back then, we only had a few owners of the motorized push net boats (*panakagan*). You work for them. They are the ones who have money in the village then. They own the boats and we work for them.[23]

Only two people ran the village then; they are the ones who had money. Now we can get many financers. Most cage producers here have financers.[24]

Aquaculture transformed villagers into individual entrepreneurs, in many ways creating new capitalist subjectivities. Yet this transformation hides forms of village cooperation and sharing. Knowledge about breeding techniques circulates freely, and strains of breeders from other parts of the country that are crossbred to improve fish growth are also commonly shared among producers. Pooling of fingerlings to supply the requirements of large fishpens has been a common practice in Kalinawan and elsewhere since the 1980s, given that pens require anywhere from one hundred thousand to a million fingerlings, and one nursery can produce only thirty to fifty thousand.

The nearby lakeside village of Navotas in the municipality of Cardona experienced the beginnings of small-scale cage aquaculture at around the same time as Kalinawan.[25] But unlike Kalinawan, where capture fisheries operations were almost completely replaced by cage nurseries, aquaculture in Navotas continues to provide only one of several lake-based livelihoods that villagers draw upon for daily subsistence. Navotas did not possess the locational advantages and ties to a research station that Kalinawan had. Until the trader-transporter (*naghahango*) emerged in the early 2000s, motorized push net boat owners remained the most affluent villagers. With the crackdown following the enforcement of the Fisheries Code, many of these boat owners converted their boats for aquaculture trading-transporting or seine harvesting of fish, which many fishpen operators outsource to villagers.[26]

In 2002, Terry became the first Navotas villager to engage in aquaculture trading-transport. With help from two financial partners and building on her previous contacts as a fish retailer and assembler (*digaton*) in the municipal fish port, she established contacts with pen operators and caretakers. Traders like her transported pen-harvested fish to lakeshore municipal fish ports and the Manila fish market. After hiring seiners (*takibo*) to corral fish, fishpen operators would contact traders like Terry to negotiate a price for the fish, which they would buy, haul to their boats, and load in trucks to sell in the Navotas fish market in Metro Manila. Terry noted the emergence and growth of trading-transporting in Navotas as many others in the village followed her lead and achieved success. This intensified competition eventually led to her income taking a hit.

Berting and Sonny, both local village officials, were elected largely due to their success in trading-transporting. From working as fishing crew in motorized push nets, they were able to save enough money to purchase their own boats and trucks for trading. They have continuously built a network of contacts with fishpen operators. Entry barriers to trading are significantly lower than in motorized push net fisheries. With access to credit, trader-transporters can earn enough to recover initial costs within one or two trips to the fish market.

The organization of fish trading-transporting has more similarities to capture fishing operations than to cage production despite relying on the fishpen economy. Labor recruitment is limited to the village, with crew composition stable over time. Additional laborers are hired if more fish are to be harvested, and the trader or the boat captain decides who to allow to work, often based on patronage and kinship ties. The share tenancy system in capture fisheries, however, disappeared with the rise of trading because crew members are paid a nominally fixed although often fluctuating wage. Income from trading is less variable and seasonal than from capture fisheries owing to the stable and controlled volume and frequency of harvests in pens.

The introduction of aquaculture involved significant transformations in the social relations and livelihoods of lake villagers. Compared to capture fisheries, aquaculture development brought varying degrees of change in how fish is produced and the institutional arrangements surrounding such processes. Cage nursery production reworked ownership of the means of production, labor arrangements, and patronage relations between villagers. A similar process occurred in fishpen-related trading, but differences from capture fisheries institutions are not as significant.

Cage nurseries depend on seed demand from pens, and the incomes of village traders depend on strong pen harvest. Pen expansion did not completely displace capture fisheries livelihoods as expected in capitalist enclosures because of fisherfolk resistance, state regulation, and the availability of migrants from other regions who provided labor for fishpens. Furthermore, cage aquaculture employs household and occasional wage labor to produce cheap inputs for pens, while capture fishers (converted fishing boats) provide specialized labor for tasks that pen laborers cannot perform. In this sense, rather than complete proletarianization, capitalist pen production indirectly appropriated noncommodified labor and nature through the work of small-scale aquaculture producers and capture fishers.

The development of small-scale aquaculture in Kalinawan and other lake villages has created new types of individualized entrepreneur-producer subjectivities—the fisherfolk as petty commodity producers and capitalist subjects (Li, 2014)—as a result of deconcentration of ownership from a few owners in traditional capture fisheries to several households in cage aquaculture. It has also led to stronger connections with pen production and further differentiation among villagers, ranging from highly capitalized bighead carp hatchery owners to cage aquaculture producers and wage laborers. The upstream and downstream dependence of cage nursery producers, village traders, and fisherfolk harvesters on pen operations (seed inputs, harvesting, and ancillary pen work) has created a complex patchwork of agrarian relations in the lake that similarly shapes how they view the legitimacy and continued presence of large-scale fishpen operations.

CONCLUSION

At Laguna Lake, the success of capitalist aquaculture lay in its ability to harness the high ecological surpluses of a newly opened commodity frontier—the gifts of plankton, migrant labor, and unpaid work of small-scale producers at the height of the fishpen rush—through enclosures. When these arrangements subsequently peaked due to exhaustion of such configurations and engendered resistance from and violent conflicts with lake villagers dispossessed of their fishing grounds, new forms of state regulation emerged that sought to manage conflicts and return profitability and productivity while extending territorial state power over a frontier in chaos.

However, the intrusion of capitalist aquaculture in the lake and subsequent agrarian transitions from fisheries to aquaculture in the villages have been spatially uneven, partial, and nonlinear. Villages such as Kalinawan adopted aquaculture more fully than others like Navotas due to place-based path dependencies and conjunctures. Aquaculture also remains one of several activities that many villagers use to make a living directly and indirectly, a form of diversification that reshaped the depeasantization and proletarianization imperatives of enclosures and created new capitalist subjectivities.

Commodity frontiers are transformed by capital in ways that are expected, such as the emergence of enclosures and associated dispossessions, contradictions, and exhaustion. But such transformations are also contingent and conjunctural, creating a patchwork of agrarian relations in which various ways of organizing fish production coexist and capital's relations with agrarian production take multiple forms. The next chapter takes these questions of conjunctures further by illustrating three encounters in which nature's materiality shapes the socioecological fate of an unruly frontier.

THREE

An Unruly Frontier

MODERN FRONTIER MAKING IS A STORY of erasures and effacing to establish a new order. Yet histories of ordering and disciplining schemes to enable delivery of vital resource flows are filled with narratives of nature failing to conform to human plans. Humans and nonhumans that populate landscapes are active participants in the making of resource frontiers despite attempts to simplify the socioecological terrains or render them invisible and extractable. This chapter extends narratives of frontier making and ecological transformations discussed in the previous two chapters by demonstrating the multiple ways that frontier lives and natures redefine the structuring trajectories of resource production. It does so by taking three views of the matter of the materiality of nature: how properties and objects of what we term "natural" play an active role in the making of histories and geographies in resource frontiers.

Nature's materiality emphasizes the unruly and lively nature of frontier making in Laguna Lake, where the unexpected creates new socionatures that do not always fit people's desires. I present three such encounters in the lake: state schemes and hazards, capital and aquaculture production in a lake environment, and invasive fish and livelihoods. Each offers accounts of nature-society constellations that reconfigure frontier trajectories beyond the initially intended.

First, hazards such as typhoons and floods regularly frustrate attempts to control fish production in the lake but have also facilitated regimes of regulation intended to manage aquaculture's excesses. Both suggest that hazards in Laguna Lake are intrinsic to, rather than merely external forces in, the history of resource frontier development. Second, capital confronts the distinct materialities of producing in the lake, which shape the trajectories of accumulation, organization of production, and relations between producers. Third, unruly invasive fish assemble new practices, performances, and communities as livelihoods are entangled with the vibrant emergence of invasive life.

These three views of materiality begin with different premises about the ontology of nature and present varied, if conflicting, understanding of nature and the nonhuman.[1] However, they are held together by the narrative ironies of surprise, contingency, and conjuncture, and by a metabolic framing of the resource frontier as a shifting assemblage of human and nonhuman work and practices.

VIEW 1: HAZARDS FOILING STATE SCHEMES

In 1970, just a few months into its operation, the LLDA experimental farm suffered a setback when a typhoon struck and damaged the newly constructed fishpen structures. Milkfish, stocked to evaluate their potential as an aquaculture species, escaped the pen enclosures. The typhoon delayed the first harvest until the following year and would foreshadow a recurring element in the history of Laguna Lake aquaculture. Typhoon and flood damage to the experimental farm recurred throughout the 1970s and was regularly documented by the LLDA's annual reports in terms of costs of structural damage and percentage of stocked fish lost.

Two years later, the great flood of July and August 1972, which inundated many parts of central and southern Luzon, doubled the lake's volume and brought the water to levels not recorded since September 1919 (Laguna Lake Development Authority, 1972; Nilo & Espinueva, 2011). It caused extensive damage to both the experimental farm and other early privately operated fishpens in the lake. The LLDA (1974) scrapped another fishpen demonstration project on the lake's East Bay because of recurrent strong waves, inundation, and typhoon damage, eventually marking the project as an extraordinary financial loss. In late 1978, a series of typhoons—Weling (Lola), Yaning (Ora), and Kading (Rita)—damaged 65 percent of stocked fingerlings, financially constraining the sustainability of the experimental farm (Laguna Lake Development Authority, 1979).

Feasibility studies and LLDA scientists very quickly recognized, even at the early stages of the experimental pen, the need to incorporate the possible effects of typhoons and floods in pen design and operations (Laguna Lake

Development Authority, 1978a). Vicente Lavides Jr., LLDA's first general manager, noted that the LLDA staff had always encountered typhoon damage and were "in continuous search for materials which could be utilized in the construction of fishpens which have longer economic life and are resistant to wave and wind action" (Laguna Lake Development Authority, 1972, p. 6). Hazards such as typhoons and floods were intrinsic to the development and trajectories of Laguna Lake aquaculture from its early origins and not merely occasional external disruptions to an otherwise stable production system applied using abstract, scientific agricultural knowledge.

Working with Hazards

The 1976 typhoon season also reduced by half the incipient commercial private pen area in the lake and temporarily halted rapid pen expansion, which had developed parallel to but separate from the experimental farm (Laguna Lake Development Authority, 1978a; Palma et al., 2005). However, these private operations were quick to recover and reconstruct despite getting little technical support from the LLDA and relying primarily on practical knowledge and financial resilience of producers. With an average of twenty typhoons entering the Philippine Area of Responsibility over the long term, six to nine of these making landfall annually since 1970, adjusting and responding to typhoons became a crucial component of agricultural production in the country (Blanc & Strobl, 2016).

The ability to incorporate typhoons into pen operations is inherently uneven, with larger-scale pens more successful at investing in technological adjustments, reflecting disparities in financial capacities. These adjustments include the use of sturdier but costlier palm trunks (*anahaw*) instead of bamboo poles and innovations in net design to reduce the threat of fish spillage. For Joel, a pen operator, typhoons and floods pose a significant problem only to those inexperienced operators who use shorter nets that would likely get inundated during a typhoon. A good operator, according to him, would also ensure that the fishpen can withstand strong winds by making sure that the poles are new and well-maintained. Because early forecasts give operators a few days to prepare, decisions about whether to harvest early to save stocked fish will depend on experience of past typhoons, the storms' behavior and effects, and current financial conditions.

The experimental farm and private pens made parallel changes in the pen design and techniques of production through experience, observation, and constant dealing with winds and waves. They developed knowledge about aquaculture and the lake through practice and encounter, combined with access to capital to recover from typhoon damage. This is in stark contrast to the state-funded aquaculture fishpen projects, such as the LDBFDP, which ended up as considerable failures because fisherfolk beneficiaries lacked prior experience and were unable to bounce back from significant financial losses after typhoons and floods. For smaller fishpen operators like Bryan, the financial risks of typhoons are magnified:

> This kind of business is a test of patience and endurance. In the lake, we say it is better for your fishpen to get poached on rather than to get damaged by a typhoon. You can lose everything you worked hard for.[2]

Fishers, for their part, have always seen floods and typhoons in a different light. To take advantage of fish that escape from damaged or overflowing pens, many fishers reinvest in gill nets that catch milkfish and other escaped fish. "Life is comfortable after typhoons," is a common claim by fisherfolk, who see these hazards as bad news for the rich and good news for the poor. Some fishers view these events as opportunities to accumulate cash and invest in better gear that will then help increase their income from subsistence fishing. Other cage producers invest in gill net gear to take advantage of abundant milkfish and reduce losses in their cage production. As one village fish assembler observed: "As long as the typhoon is not that strong but enough to damage fishpens, people here, the ordinary fisherfolk, have money. During any other time, they just barely get by."[3]

In this situation timing is important. On the day after a typhoon, the escaped fish in the lake are plentiful but not enough to produce a glut in the fish market that would depress prices. Milkfish can drop to a tenth of its average nontyphoon price. Village fish traders work quickly to bring the fish to the urban wholesale fish market to capitalize on the temporarily high difference in fish prices between those at the lake and in the city. Producers or traders who have greater storage capacity for fish are able to buy fish cheaply from the lake and sell them with considerable markup in the urban fish market once the initial glut subsides. However, once the post-typhoon dash for escaped fish has waned, hauling of fish from pens becomes less frequent, as it takes time for pens to be stocked again and harvested.

Production is a balance between strategy (*diskarte*) and pushing one's luck, similar to gambling (*pakikipagsapalaran* and *pagsusugal*). The impacts of typhoons and floods on lake ecologies are not a given and are variable and complex. Villagers note that Tropical Storm Ondoy (Ketsana) in 2009 brought good productive conditions for the next several months after affecting the lake, whereas Typhoon Basyang (Conson) achieved the opposite effect the following year. Producing in the lake therefore involves constant adjustment that becomes the basis and opportunity for deepening fishers' practical knowledge. In the face of hazards, producers deploy practical knowledge to avert, mitigate, or prepare for damage, thereby stabilizing aquaculture production as an assemblage of varying human and nonhuman elements. These assemblages break down with disruptions in particular elements, such as an error in the weather forecast or specific changes in producer ability to prepare or recover.

Hazards are often seen as either a problem or an opportunity. A thin line, however, separates typhoons and floods as hazard/risk or as resource/benefit for agricultural producers (Bankoff, 2003; Eakin & Appendini, 2008). In Laguna Lake, cage and pen producers perceive typhoons and floods as hazard/risk given that aquaculture fixity exposes production to damage, while fisherfolk see them as resource/benefit in view of the opportunities they present to improve their incomes. In contrast to the narrow state optic of control, efficiency, and productivity through application of modern scientific knowledge, producers combine various elements that are useful for them through observation, experimentation, and sharing with other producers (Scott, 1998).

In contrast to the notion of hazards as disruptions to normal life or to an otherwise stable production order (Bankoff, 2003), producers continue to live with hazards, internalizing them as intrinsic to the assemblage of fish production. The notion of living with hazards is different from the modern separation of society and nature, often deployed by experts, technocrats, and state actors who consider hazards as external forces and frame them as problems to be solved or objects to be controlled (Scott, 1998; Mitchell, 2002). Rather, hazards such as typhoons and floods are among the elements that are part of, and that shape the architecture of, aquacultural production.

Frustrating Regulation

Hazards also shaped the trajectory of the state's regulation of aquaculture through the ZOMAP, first formulated in 1983 (see map 3 in chapter 2). Despite constant efforts to manage aquaculture sprawl, dismantle illegal fishpens, and order the lake through the proposed geometric belts, implementation of ZOMAP took thirteen years to materialize. Annual reports from 1983 onward carried promising accounts of ultimately futile efforts at fishpen demolition as a result of various strategies employed by fishpens to circumvent regulation and the difficulty and physical labor required to actually dismantle pens. The work required to demolish fences and enclosures is as significant and painstaking as the work of constructing and maintaining them.

The planned modern restructuring of lake space through the ZOMAP eventually took place in the 1990s through a convergence of the unintended and the more-than-human. On November 3, 1995, the very strong category 5 Typhoon Rosing (Angela) hit the Laguna Lake region. The LLDA had formulated the Laguna de Bay Master Plan, approved by President Fidel Ramos a month before the typhoon, which further limited the total fishpen area to 10,000 hectares. The LLDA had also intensified the demolition of pen structures despite continued jurisdictional conflict with the local governments and resistance from pen operators (Carlos, 1995a; Laguna Lake Development Authority, 1995a; Tribdino, 1996). Ramos framed these interventions during his presidency (1992–1998) within environmental protection, sustainability, and pollution management (Tribdino, 1995; Carlos, 1995b) in the context of his administration's vision of the Philippines emerging as a "green tiger" (Goldoftas, 2006) that introduced neoliberal innovations in environmental governance (Oledan, 2001).

Nearly all of the fishpen structures in the lake were damaged by Typhoon Rosing, which, when combined with the destruction caused by Typhoon Mameng (Sibyl) in 1995 and Typhoon Katring (Teresa) in 1994, severely affected the ability of pen operators to recover (Carlos, 1994, 1995b; Laguna Lake Development Authority, 1995a; Palma et al., 2005). Giving credit to Typhoon Rosing for doing the work of clearing pen structures in the lake, Environment Secretary Victor Ramos issued in November a moratorium on pen repair (Carlos, 1995c; Laguna Lake Development Authority, 1995a). The moratorium gave the LLDA enough time to institute the ZOMAP, which it adjusted and enforced in 1996.[4] The LLDA needed what the environment secretary referred to as unplanned acts of divine providence and natural forces to finally institute the highly modern (and very human) planning project of zoning the lake (Carlos, 1995b, 1995c).

By placing hazards more centrally in modern state development schemes, typhoons and floods may be viewed as forces that either disrupt or support state visions of improvement and frontier making. Typhoons, for example, frustrated or reconfigured attempts to establish aquaculture enclosures in the lake. Typhoons became factors that state scientists and pen operators needed to overcome in order to institute state schemes and make them successful. The same hazards also presented opportunities for state managers to enforce the LLDA's zoning plan as a solution to the problems enclosures brought to the lake. The LLDA's character as a state institution was strengthened by a confluence of human and nonhuman forces, including stronger national government support, the Supreme Court ruling reaffirming LLDA's jurisdictional mandate, and a series of typhoons that damaged existing aquaculture structures. However, this binary—or perhaps spectrum—of hazards as constraint or resource may be considered fundamentally intrinsic to these schemes.

Hazards such as typhoons and floods become central protagonists not only as external objects that shape social actors or forces that they respond to but also as something internal to how these inherently hybrid socionatural modern development schemes emerge historically. The fates of frontiers and their regulation are shaped by the forging of alignments and coming together of various elements in an assemblage, human and more-than-human, planned and beyond intended (Li, 2007; Mitchell, 2002).

View 2: Materialities of Capitalist Aquaculture Production

Capital confronts nature in its spatial expansion and deepening intrusion into metabolic relations of production. In places like Laguna Lake, this process permeates the making and maintaining of a commodity frontier. As a method of producing fish that seeks to transcend the limitations of capture fisheries, aquaculture is itself a sociotechnical innovation that continues to develop with greater control and intensification of production to extract or appropriate more surpluses in frontiers opened up for commodification. Working with nature's materiality, in this sense, produces particular ecologies as capitalist aquaculture employs various strategies that range from improvements in production techniques and changes in property rights to genetic improvements of the fish itself. In the language of critical political economy, this signifies capital's formal to real subsumption of nature (Boyd et al., 2001). Central to the logic of capitalism itself, nature—from bodily to biophysical—becomes an obstacle, opportunity, or surprise (Benton, 1989; Boyd et al., 2001; Goodman et al., 1987; Guthman, 2011; Henderson, 1999; Kloppenburg, 2005) coproduced through labor in metabolism (Eaton, 2011; Ekers & Loftus, 2013; Smith, 2008).

Production in commodity frontiers such as Laguna Lake continues to confront biophysical nature in ways distinct from capture fisheries or agriculture based on land. While land is often considered immovable and fixed, water is characterized by properties of fluidity and circulation, with implications for commodifying water (Bakker, 2004) and producing commodities in bodies of water (Mansfield, 2004; Campling, 2012; Sneddon & Fox, 2012). Water bodies also serve as sinks for effluents from surrounding activities, and waste by-products tend to undermine the conditions necessary for sustained production and profitability. These diverse ecological qualities of agrarian space contrast with capital's attempt to homogenize and render equivalent the natural world.

Aquaculture producers have continued to employ various strategies to improve production and work with particular lake natures, with varying degrees of success. These ecologies of production have involved the introduction of new production techniques and methods of organization of production that are unique to the conditions of aquaculture and that differ from and seek to improve on capture fisheries. Laguna Lake aquaculture is distinct, relying on the natural food abundant in the eutrophic lake, the urban edge location of which highlights the multiple conflicting production of ecologies. Various materialities influence the possibility of continuous production, the species reared, the timing of production tasks, the ability to intensify production through increased stocking or feeds, the deployment of labor, and inter-producer relations.

Contradictions of Saltwater Intrusion

As chapter 1 showed, fish producers and scientific assessments both attest to the importance in production of the seasonal incursion of saline water into Laguna Lake via the Pasig River. Occurring during the dry months of April and May when the lake water level begins to fall below sea level, the circulation of saltwater helps clear the turbid lake, thereby improving photosynthetic activity and increasing primary productivity, or the conversion of abiotic components of the lake into biotic through photosynthesis. This is usually followed by increased abundance of phytoplankton and zooplankton in the lake, which then improves fish growth by providing food and reducing pathogenic microorganisms (Palisoc, 1988; Santiago, 1990; Santos-Borja, 1994).

Saltwater intrusion speeds up production time in fish culture and allows a faster turnover whenever two or three crop cycles in a year become possible.

During poor water conditions, it would take up to a year or more for fish to grow to marketable size, a situation that producers would refer to as fish "celebrating their birthdays." Both pen and cage producers attest to the ecological significance of this intrusion:

> What we need here is seawater. With saltwater intrusion, the lake clears, and when the lake clears, it provides more food. Back when we had intrusion, cage producers were able to harvest three times a year. Every three months they were able to harvest tilapia. These days, the fish already celebrated two birthdays, yet they are still this small.[5]

> In the fish culture of Laguna Lake, saltwater intrusion is vital because that is when the fingerlings are stocked during the dry season. Fingerlings adapt better to the lake conditions with salinity because it takes milkfish so many days to adapt. Fingerlings also survive better. It also has a cleansing effect. Salinity kills parasites, and so milkfish fingerlings survive better. Without saltwater, like this year, many fishpen operators are not doing so well. The only ones who survive are the well-off ones because they just keep on stocking even with low survival rates. With saltwater in the Taguig-Napindan area where my fishpen is located, our survival rate is about 65 percent. If there is none, we have 10, 15, 20 percent survival rate.[6]

The coincidence of shorter production times and the deployment of labor may help explain why pen production was very profitable during the height of the pen sprawl in the late 1970s to early 1980s, when saltwater intrusion was as yet unaffected by the saline backflow control of the hydraulic control structure. However, saline incursion is not without drawbacks, since the same flux poses risks to fish health. Passing through the Pasig, a polluted river that cuts across Metro Manila, flows of saline water are also accompanied by excessive nutrient fluxes that can cause sudden fish kills, one of the points of contention in debates about hydraulic control of saline flows mentioned in chapter 1 (National Statistical Coordinating Board, 1999; SOGREAH, 1991). The flux of saline water is spatially and temporally uneven, with areas of the West Bay closest to the river receiving it first and most, making them the most productive areas but also potentially the most threatened with fish mortality (Cendana & Mane, 1937; National Statistical Coordinating Board, 1999; Villadolid, 1933). Joel, a mid-sized pen operator who had experience in losing stocks to fish kills, for example, referred to his location near the junction of the river and the lake as both a blessing (*grasya*) and a curse (*disgrasya*).

The fluctuating production times associated with poor water conditions enable a host of responses in production organization and relations in both fishpens and fishcages. Supplemental feeds are given to fish in the hopes of speeding up their growth to marketable size. However, while they can increase the live weight of fish, the costs of artificial feeding are high enough to make continued reliance on them unattractive for producers. Both cage and pen producers comment that the marginal benefits of faster growth of fish through supplemental feeding are countered by the high cost of feeds. The provision of supplemental feeds also encounters the materiality of the lake—the contrast between water's fluidity and aquaculture's fixity—wherein feeds given in one fishpen end up in another due to the lake's water circulation.

Another response to fluctuating production times is to stock fish species that grow better in poor water conditions. Pens shift from milkfish monoculture to bighead carp mono/polyculture because operators have observed that the latter grow better in turbid and less productive periods when desired species tend to celebrate birthdays. Despite the carp's fetching much lower prices in the urban market, bighead carp culture allows pens to circumvent the long production time of milkfish. In pens, a long turnover time due to poor water conditions entails more work for hired caretakers and laborers.

Despite the nonoperation of the saltwater control function of the hydraulic control structure, producers still experience years without saltwater intrusion. Producers and ecologists have proposed a few reasons. Water levels do not subside to below sea level, and siltation from surrounding agricultural, industrial, and domestic activities has caused a shallowing of certain parts of the lake, which prohibits adequate seawater flux. Increased nutrient loading may have pushed the lake to a hypereutrophic state such that saline intrusion has marginal and minimal impacts on primary productivity (Zafaralla et al., 2005). These impacts underscore the ecological contradictions of conflicting demands on the lake and the multiple production of lake natures by activities both within and beyond the lake.

The Problem with Plankton

Laguna Lake's high-nutrient character enables higher primary productivity whenever the turbid lake clears and allows more photosynthesis to take place. This abundance of nutrients, state-commissioned studies in the 1970s and 1980s believed, could be more efficiently utilized through the introduction of aquaculture and subsequent innovations in production techniques.

The lake's eutrophic character has allowed fishpen operators to produce fish that are comparatively cheaper than similar fish species reared in places where artificial feeds comprise a significant bulk of production costs. Plankton abundance has been seen by fishpen operators as a way to profitably produce fish by reducing costs of inputs and supplying fish at a cheaper price. In the case of milkfish, producers in other areas of Luzon, such as Pangasinan, Bulacan, and Taal Lake, spend more than half of total costs on commercial feeds. As a result, Laguna Lake pens can supply milkfish with prices up to 25 percent lower (Tan et al., 2010). According to a fishpen operator:

> In the lake, we do not feed the fish, everything is natural. It will depend on plankton growth. First of all, you cannot do feeding. If you intensify through feeding, other fish and birds will get to the feeds first. You cannot feed, unlike in Pangasinan where they have full feeding because they do not have natural food. When the water and weather were good many years back, Pangasinan producers would stop harvesting whenever we started harvesting here in Laguna Lake because they could not compete with our low prices. What they would do instead to compete was to harvest bigger fish.[7]

Dependence on natural plankton in the lake instead of artificial feeds, however, poses a few problems in aquaculture production. It creates a high reliance on lake processes such as saltwater intrusion, increasing operator vulnerability to fluctuating production times. Although reports have indicated that pens produced two or three crops per year in the late 1970s prior to the fishpen sprawl, various factors contributed to the decline in plankton abundance, making it almost impossible to sustain this level of productivity (Richter, 2001). Dependence also limits the potential for intensification by supplementing feeds because it is not only economically unattractive but also leads to added nutrient inputs when feeds are not fully consumed by the fish. Additional nutrient load in the lake provides a fertile ground for algae blooms, which can cause mass fish mortality when they decay. It can also contribute to increased levels of eutrophication, which can contradictorily reduce primary production (Tamayo-Zafaralla et al., 2002; Zafaralla et al., 2005). Water's capacity to circulate also implies that feeds do not necessarily stay within the area for which they are intended, and it is common for neighboring pens to benefit from feeds distributed in another (Garcia & Medina, 1987; Richter, 2001).

Control of the market size of the fish is less predictable than for those fully fed on artificial feeds. This implies that fish cannot be grown to larger sizes, imposing constraints on the attractiveness of producing for the export market, where fish of bigger and more consistent sizes are desired. In less productive years, pens harvest even before the fish reach market size, to hasten the turnover cycle and limit nonproduction time. The varying sizes of fish produced in the lake find a market with poorer urban consumers (see chapter 5). Fish also acquire an unpalatable earthy off-taste (*lasang gilik*) whenever blue-green algae blooms (*Microcystis* spp.) occur during the transition from the dry to the wet season. This makes it almost impossible to sell the fish because consumers are repelled, and traders refuse to buy the fish. A trader-transporter emphasized this difficulty and his ability to affectively respond to such situations:

> You cannot harvest fish that smells of *gilik*, it would be like selling shit; no one will buy from you, not even fishball processors. That is why before I harvest from pens, I smell the fish first, sometimes I would fry one piece to make sure there is no smell. ... So once I get to the pens and notice the smell, I won't buy the fish even if they sold it to me for P1 a kilo because no one will buy them from me in the wholesale fish market.[8]

Pen and cage operators work with these materialities of production by adjusting the timing of harvests to avoid the off-taste. The staggered harvesting in pens, however, enables them to adjust better than those using cages to the disruption in production time and the turnover of a new cycle of production. More importantly, the dependence on lake plankton for fish feed is at the center of the difficulty of large, capitalized pens to further intensify and speed up production.

Fish as Lively Fugitive Commodities

Since fish can only be stocked at certain densities without raising mortality rates or affecting overall production time, the tendency is for pen operations to expand in size. Taking advantage of economies of scale, producers can cut down on the costs of materials for construction and operation, as well as on labor costs (Israel, 2007; Tan et al., 2010). However, expansion of size encounters problems with state regulation concerning the 50 ha maximum allowable pen size. With more stringent state regulation following the ZOMAP, pens were able to circumvent the size problem by creating several dummy corporations that enabled them to informally rear fish in hundreds or even thousands of hectares of lake space.

FIGURE 6. Main house inside a fishpen, with a view of the city in the background, 2012. Photo by author.

Fishpen owners hire a caretaker or administrator (*katiwala*) who is placed in charge of daily production operations involving waged laborers (*tauhan*), who are hired to do all-around work such as stocking pens, repairing nets and fences, surveillance, and other daily maintenance work. Fifty-hectare pens would employ an average of around fourteen laborers, who often come from outside the lake area, usually from the poorer central Philippine regions of Bicol and the Visayas. The aversion of local lake people to direct employment in pens is primarily due to the low pay, with monthly wages being P3,000, less than half the minimum for the area, but also because it ties them to full-time work in the middle of the lake that leaves no room for engaging in other livelihood opportunities. Pen operators view lake villagers as lazy, untrustworthy, and not amenable to labor discipline in pens, whereas migrant workers with no roots in and connections to the lake area are less likely to complain and cheat or steal. While wage laborers live in the main houses or the guardhouses in the fishpens throughout the year (see figure 6), fishpen operations also rely on seasonal hired labor from the villages for preharvesting activities such as seining.

The large pen size and scale of operations make it difficult to monitor fish. A significant amount of labor time goes into surveillance to prevent escaping fish and poaching in the pen. This is done through regular monitoring by pen laborers of the conditions of nets through "visiting" (*pagbibisita*), the construction of guardhouses where laborers are stationed, and nightly monitoring with the use of searchlights. Poaching is a common occurrence in both pens and cages, due to both the difficulty in surveillance of large parts of the pen and the ease with which poachers can swim to the pens, cut nets underwater, and set their gill nets to trap escaping fish.

Surveillance is a crucial component of labor in pens, and it occupies a significant amount of labor time. Pens have organized various strategies of improving efficiency in surveillance, which include management of labor spatially (distributing them throughout the pens in various guardhouses) and temporally (working in shifts throughout the twenty-four hours). In the early days of fishpens in the lake, this had been a common way of organizing operations, which used a variety of methods, technologies, and systems to coordinate surveillance among workers (Dela Cruz, 1982).

The relative remoteness of pens—located in the middle and least accessible parts of the lake—and the spatial noncontiguity of laborers distributed in various guardhouses stress the importance of trust relations between operators, caretakers, and laborers. Absentee operators, often based in Metro Manila, rely on caretakers who manage the daily operations of the pen. The caretakers, meanwhile, need to ensure that the laborers are performing their tasks and that they are not stealing (or allowing others to steal) fish from the pens. Bryan, a pen operator, described the hierarchy of pens thus: "I will call my main caretaker, and he will then call guardhouse caretakers and they will then relay orders to the laborers."[9] This mirrors relations in land-based fishpond aquaculture elsewhere in the Philippines (Dannhaeuser, 1986) but is complicated by the size and isolation of operations.

A history of violence has accompanied pen operations in Laguna Lake owing to the scale of their operations, which displaced traditional fishing grounds of capture fisherfolk; their use of armed guards; and their limited and largely impersonal interaction with surrounding fisherfolk villages (see chapter 2). While cage producers closer to the shore have also experienced poaching, it is often viewed differently in light of the strength and character of the social relations within the community. A lack of *pakikisama* (fellowship) and *inggit* (envy) is given as a reason for fish being stolen by neighbors. Nonetheless, surveillance is required for both pens and cages, especially close to harvest time. As with large pens, cage aquaculture requires close monitoring and surveillance. The need for regular monitoring is tied to the poor visibility of fish as fugitive commodities underwater and as a mobile resource fixed in space by aquaculture.

*Materiality of Nature and the Continuing
Persistence of Capitalist Aquaculture*

Laguna Lake aquaculture has plenty of "natural" obstacles to capital, including its extensive and semi-intensive character and highly fluctuating production times. Opportunities to overcome these obstacles by capitalist investment in other aspects of agriculture, such as seeds and feeds, have been quite limited. The improved seeds used in the lake, developed by public research institutions, have made only insignificant contributions to improving production time because the water conditions continue to be suboptimal. The same has been true for feeds; efforts of agribusiness to create a feed market have been constrained by the ineffectiveness of artificial feeds in speeding up growth of fish.

It appears that the materialities of Laguna Lake aquaculture should make it unattractive to capitalists, and thus production could be left in the hands of small-scale producers. Indeed, intensification has been thwarted primarily by the lake's eutrophic character. However, despite decades of state regulation, fluctuating limnological conditions, and the undermining of their own conditions of production, pen operations continue to dominate lake fisheries rather than leaving aquaculture to small-scale, household-based cage production.

Explanations may be found in strategies to continue to mobilize abundant cheap labor and natural plankton on a resource frontier. Early debates in agrarian political economy posed questions about capitalist agriculture's dependence on natural conditions and its implications for the coexistence and survival of noncapitalist producers amid capitalist penetration in agriculture (Mann & Dickinson, 1978). Natural obstacles constrain the use of wage labor in the countryside because of the nonidentity of the relatively fixed biological nature of production time and (human) labor time in agricultural production. Capitalization of agriculture progresses quickest in those spheres where these times overlap.

The availability of migrant labor, paid very low wages, allows pen operations to remain profitable in times of slow production. Payment through monthly wages that are often well below the minimum, supplemented by a share of the harvest, enables flexibility on the part of pen operators during extended production cycles due to poor water conditions. Also, wage labor is more flexibly employed throughout the year to attempt to match the materiality of the fluctuating biological production time of fish, given that laborers work on all types of tasks in pens (Mann, 1990). The degree and nature of the work are also adjusted depending on the water conditions and fish growth. Since laborers live in pens in the middle of the lake, they are always on call, as the work of maintenance and surveillance is a continuous process. The labor patterns therefore overlap with, although they are not determined by, the biophysical character and timing of production (Das, 2014).

Dependence on plankton, though a constraint on intensification, keeps production costs down during times of good water conditions. It allows continued production of cheaper fish sold in the urban market compared to those that use feeds. Some pen operators shift their species of choice from milkfish to bighead carp to be able to reduce production time and speed up turnover. Labor and plankton, combined with the political power of pen operators and increasing demand for cheaper fish for the city, have enabled the continued existence of capitalist pen aquaculture. Strategies of tapping the gifts of labor and plankton, while not without their problems, have proved advantageous compared to aquaculture elsewhere and to capture fisheries, even without further attempts to intensify or introduce new techniques.

Cages, on the other hand, coexist with pens by producing a different species (tilapia) that is destined for urban consumption in much smaller volumes. Fish in cages remain an important source of food and serve as a "bank in the water" (Bene et al., 2009) during tough times. Cage production is a relatively profitable livelihood option for village fishers able to overcome its barriers to entry. Cage production and capture fisheries are linked with pen operations in various ways, reflecting the changing relations between capitalist and noncapitalist aquaculture in Laguna Lake. Thus, apart from plankton, cage producers' labor power reproduced outside of capitalist relations may be seen as being appropriated by capitalist aquaculture. Despite capital's attempt to subsume nature completely and commodify it, noncapitalist formations are needed to continue its reproduction. Nature therefore matters in particular ways to the agrarian question of capitalist aquaculture and its implications for labor and agrarian relations in commodity frontiers.

VIEW 3: ENTANGLED INVASIVE LIVES

Aquaculture as a modern intentional intervention made it possible for unintended, introduced species to transform host socioecologies. In Laguna Lake, introduced aquaculture species have come to dominate total fish production, as well as catch composition of capture fisheries at the expense of indigenous fish (see figure 3 in chapter 1). Like typhoons, plankton, and saltwater intrusion, the proliferation of unwanted invasives has fundamentally reorganized lake production relations and state regulation. Two fish—the hardy janitor fish and the predatory knifefish—have brought dramatic changes to fisheries and lake livelihoods.

Invasives have social lives with their own geographies, as pointed out by nature-society scholars (Everts, 2015; Frawley & McCalman, 2014; Robbins, 2004). Following this body of work, the third example of nature's materiality in this chapter tracks how relations, communities, and networks emerge in response to invasive life. It emphasizes entanglements that produce invasive lives, the assemblage of which is formed through multiple encounters between humans and nonhumans (Head & Atchison, 2015). Narratives of such encounters (Faier & Rofel, 2014) show the processes of embedding, reassembling, and coproducing invasive life that remake environmental practices and subjectivities in the lake.

Entangled Livelihoods

Janitor fish or sailfin catfish (*Pterygoplichthys disjunctivus* and *P. pardalis*) began posing problems for Laguna Lake fisherfolk in the early 2000s. Colloquially named after their ability to "clean" aquariums, the ornamental fish indigenous to the Amazon have been accidentally released by aquarium owners into urban waterways (Chavez et al., 2006b). They have since established populations in the Pasig-Marikina river system and certain parts of Laguna Lake, thriving in even the most polluted environments.

Fishers consider the janitor fish an inedible pest. One describes it as "a fish that you could neither sell nor eat."[10] Almost all kinds of fish, crustaceans, and mollusks in the lake, especially introduced species, have been put to use by residents as human food or animal feed. Despite attempts to utilize the fish bodies as fishmeal and other products, the janitor fish remains one that lake residents cannot consume in any form.

Apart from the fish not tasting good, the state discourages its consumption because of its ability to bioaccumulate low concentrations of heavy metals and possible contamination with *E. coli* (Chavez et al., 2006a; Guerrero, 2014). Its armor-like exterior has also been associated with its hardiness. The janitor fish tolerates poor water conditions and consumes any food available as an omnivore, making it highly adaptable to the transformed ecologies of Laguna Lake and nearby rivers (Chavez et al., 2006a).

Despite its proliferation, the janitor fish has not been documented as a major threat to other fish species. However, the fish, with its sharp scales likened by lake dwellers to a broken tin can, often becomes entangled with gill nets set by capture fishers. Livelihood impacts can range from a minor inconvenience to significant damage to fishing gear. Fisherfolk note the irony of the intended aim of the fish's introduction considering its eventual effects:

> They said the fish is supposed to clean the lake but it created more problems. Our nets always get damaged when they entangle with the fish. When they get stuck in the nets, they are difficult to remove without further damaging your net.[11]

Janitor fish also get caught in fish corrals set by fishers to passively trap fish. Instead of harvesting edible fish, fishers would often find a significant portion of their catch consisting of janitor fish, according to one estimate anywhere between 10 and 38 percent of the total catch (Chavez et al., 2006a). Fishers would destroy them because they found no use for the fish and to prevent their escaping back into the lake: "What we do is we kill them. We chop them with a machete."[12]

The carnivorous knifefish (*Chitala ornata*), which first appeared in noticeable volumes in 2011, is considered more of a scourge for both aquaculture and capture fishers alike. Its voracious appetite and reproductive behavior enable the fish to grow bigger and multiply in a shorter amount of time. Its ability to escape many types of fishing gear also makes it difficult to control. Like the janitor fish, the knifefish was either accidentally or deliberately released into waterways that drain into the lake (Abarra et al., 2017; Castro et al., 2018) sometime between 2004 and 2009. While producers report encounters with the fish in the mid-2000s, the massive floods that hit Metro Manila in 2009 and 2010 are often considered the a benchmark for its invasion.

Also colloquially known as *arowana* owing to its resemblance to another popular aquarium fish, the knifefish is described by fisherfolk based on its sizeable, elongated body and monstrous mouth that can encompass even the largest fish in the lake (see figure 7). It can grow to up to 5 to 10 kilograms in weight and inspires a sense of fear among producers and fish alike: "Knifefish is huge. Its mouth is big. It can even eat bighead carp" and "The mouth is so big, and it has plenty of teeth, sharp teeth. Even the tilapia are afraid of them."

FIGURE 7. Knifefish next to a bighead carp. Photo by author

A 2014 survey found knifefish comprised 40 percent of capture fisheries' catch, even surpassing desired species like tilapia and milkfish (Palma, 2015). Fishers also point to further declines in already dwindling populations of indigenous species such as white goby and silver perch as a result of knifefish proliferation. By 2012, fishers and aquaculture producers were raising the alarm about a potential fisheries crisis in the lake.

Encounters in Enclosures

Knifefish posed a significant threat to the large-scale fishpen operations. A kilogram of knifefish growth in weight is equivalent to 7 kilograms of consumed lake fish. The sheltered waters of fishpens and the bamboo and palm poles have become attractive sites for knifefish to feed and breed. The fish reproduces naturally, quickly, and abundantly, unlike milkfish, bighead carp, and tilapia, which rely on artificial methods of reproduction.[13] The abundance of millions of fry and fingerling inside an enclosure means that the knifefish could enjoy a feeding frenzy.

There are differential impacts on pens and cages. In cages, the presence of knifefish adds to labor requirements by forcing producers to regularly haul in nets to remove knifefish fingerlings before they reach adult size and begin preying on fish stocked in the cage. In pens, however, this is not possible because a significant population of knifefish can escape seining. Several adult knifefish can consume up to 90 percent of fry and fingerlings stocked in a pen:

> Knifefish is our problem. Unlike typhoons, which occur once every few years, knifefish is here throughout the year. They can eat seven or eight pieces of milkfish fingerlings a day. If you compute it, in two months, knifefish can consume the equivalent of a hectare's worth of fingerling stock.[14]

To reduce mortality, pen operators respond to the threat of knifefish by stocking bigger but more expensive fingerlings that are more likely to escape predation. The proliferation of knifefish within pens has also forged new forms of pen-fisherfolk cooperation, which runs counter to the historically adversarial character of their relations. Fishers have found that using trawl line (*kitang*), a traditional fishing method that had become uncommon due to declines in target indigenous fish, is the most effective technique for catching knifefish. *Kitang* is a highly selective and low-intensity type of gear that requires significant time and labor to prepare and set, often several hours more than other fishing gear. Before stocking, pen operators would allow fishers to set trawl lines within the pens, pay them a couple of thousand pesos, and allow them to keep all of the knifefish catch to sell to traders for the urban market. After the floods of 2012 that inundated the lake for several months, *kitang* became a source of livelihood for many affected fisherfolk, who would pool their lines to catch more fish.

While the knifefish has had a profound impact on the indigenous and introduced fish catch of fisherfolk, it has also created an opportunity to shift techniques and strategies in fishing. As a fish with white meat, it has been transformed into a commodity by fishers selling to traders, who then bring the fish to the urban market for processing into fish balls. It has become an even cheaper white fish alternative to farmed bighead carp, which originally replaced a more expensive marine fish (see chapter 5).

"Knifefish is unlike the janitor fish, which we cannot find a use for," remarked Nelson in 2012. "It's a pest, yes, but it helps our livelihood."[15] The knifefish invasion created new, unexpected forms of relations between pens, fishers, and traders.

Edible Invasives

As an invasive pest, the knifefish has become entangled with fisherfolk livelihoods and fishpen operations, threatening the lifeblood of fish production in the lake. State response has involved strategies of getting rid of the fish in the lake through various means. Because the knifefish is difficult to catch using most fishing gear, state fisheries scientists have experimented with methods to destroy the eggs and prevent further reproduction, including electrocuting and burying the fish eggs. The state, through an interagency working group, also mobilized a massive retrieval program wherein fishers are paid P20 per kilo of knifefish that they catch. Beneficiaries of the state's flagship conditional cash transfer program were also involved through a cash-for-work scheme involving more than five thousand lakeshore beneficiaries. Together these two mechanisms collected a total of 160 tons of invasive fish just within two years (Palma, 2015).

The LLDA's general manager saw the problem of edibility and consumption as a key component of managing the knifefish invasion: "We can turn a potential crisis into an opportunity.... The problem is the familiarity issue. They don't know that knifefish is safe to eat and has a lot of uses" (Calleja, 2012). The LLDA administration was keen on turning "crisis" into "opportunity" by playing up the economic potentials of the fish and thereby solving the problem of invasion while simultaneously providing jobs to the fisherfolk affected by its proliferation. This included training fishers to process the fish into value-added products such as hotdogs, fish balls, dumplings, and fishmeal. In 2015, a facility for processing knifefish run by a women's cooperative opened in Pila, Laguna (Cinco, 2015).

The state had similarly promoted human consumption of knifefish as a way to reduce its population in the lake. Despite its being edible, unlike the janitor fish, lake dwellers have been apprehensive about consuming the fish, which the LLDA has attributed to unfamiliarity with the fish and its depiction in the popular imagination as a foreign, monstrous, predatory creature. State officials would assure its edibility and highlight its status as a local favorite in its native Mekong River basin.

The state's knifefish eradication working group also promoted various culinary suggestions to demonstrate the multiple ways that the fish could be consumed by transforming the unfamiliar fish body into a generic white fish or identifiable processed meat that could be prepared using common, everyday cooking methods. During its fiesta celebrations, the lakeside town of Siniloan hosted a contest to illustrate the culinary possibilities of the fish. The local government organizers were eager to frame the issue as a necessity for transforming perceptions about the fish by incorporating it in local culinary imaginations and transcending the material constraints of the fish body's characteristics (Cinco, 2014).

Hybrid Invasive Lives

Knifefish and janitor fish thrive in the heavily transformed socioecologies of the lake as unintended lives that exceed state desires for control of nature. Like the mushrooms at the end of the world in Anna Tsing's account (2015), they embody the kinds of life emerging in capitalist ruins as exploited and exhausted resource frontiers in the Anthropocene. Yet fish as hybrid entities—indivisibly of nature and anthropogenic—escape and subvert the state partitioning of socionatures (Robbins, 2001). They also represent the failure of attempts to control lake nature as a modern urban resource frontier. The knifefish invasion was unplanned and unintended and threatened the ordering of the lake as an efficient producer of fish for the city.

The very condition that enabled modern aquaculture—the enclosure of fugitive fish—sowed the seeds for invasion and the undermining of its own productivity. The fish became embedded within existing networks of urban metabolisms through encounters with humans and nonhumans. The state delegated responsibility for getting rid of invasives to fisherfolk by tapping their labor and their fishing and ecological knowledge of the lake to manage a novel menace. But knifefish also transformed the constitution of such metabolic flows through new ways of knowing nature and responding to emerging socioecological conditions.

The story of the knifefish demonstrates a reconfiguration of lake socioecological relations in response to its threat, enabling intersections of mutually beneficial fisherfolk-fishpen interests, and altering livelihood opportunities and threats for capture fishers. It also enabled a reassembling of the fish body, disentangled from its unfamiliar form associated with a carnivorous predator into a disembodied ingredient for a variety of processed foods and cuisines. New relations and communities also emerged surrounding attempts to get rid of the fish through particular labor arrangements and to make the fish edible and palatable for consumption. These practices of making do and improvisations are situated within and inseparable from the constraints of transformed lake socioecologies, frontier making, and urban metabolism.

CONCLUSION

These three examples of what may be collectively called the materiality of nature point to the lively and emergent production and coproduction of resource frontiers. The three cases present a foil to the frontier-making project of regulation, privatization, and accumulation. Hazards frustrate orderly and legitimated production, lake conditions prevent deepening innovations in capitalist aquaculture production, and the invasive fish threatens commodities and communities. These materialities reshape the organization and work of state, capital, and livelihoods in the lake and redefine or resist trajectories of frontier making and resource production.

The frontier is lively, filled with diverse lives, liveliness, and livelihood making. Subverting, frustrating, and shifting intentions and plans of control, diverse entities such as hazards, plankton, saltwater intrusion, invasive fish, and many others share the narrative plane in the histories and geographies of frontiers with the usual human suspects. Through such a lens, the lake's socio-ecologies become a frontier assemblage of humans, nonhumans, and their emergent relations, shaped by capital but also by many more forces beyond it (Cons & Eilenberg, 2019). Enrolled in urbanization through metabolism, these lively materialities present significant implications for resource production and consumption beyond the lake as an urban frontier assemblage. The next three chapters therefore turn more explicitly to the sociomaterial lives embedded in flows of fish and floodwater between lake and city.

PART TWO

The Work of Urban Metabolic Flows

FOUR

Chains of Urban Provisioning

EVERY DAY, nearly 500 metric tons of fresh fish pass through the Navotas Fish Port Complex, Manila's busiest and largest fish market, on their way to urban consumers. All kinds of fish are exchanged in the Navotas market, but standard staples on Filipino plates, such as round scad or *galunggong* and other marine wild fish, have traditionally dominated landing volumes. But with declining marine catch and rising marine fish prices in recent decades, cheaper and more abundant farmed freshwater fish have taken over a significant share of fish unloading and urban fish consumption. Laguna Lake currently supplies Manila with two-thirds of its freshwater fish consumption, which is not surprising given the long history of state intervention in its ecology to enable the lake to produce more fish for the city.

Decades of urban frontier making have transformed Laguna Lake to supply much of the freshwater fish requirements of Manila and beyond. How much and where such enduring flows of fish go can be traced as urban metabolic material flows of food to the city. Following industrial ecological notions of urban metabolism, these flows may be viewed as an exchange of nutrients and energy, as fish embodying nutrients in the lake consumed by city dwellers. But as I aim to show in this chapter, fish travel as commodity flows that are also thoroughly social and worked. Fish as circulating commodities are material objects produced through lake nutrients and consumed for sustenance but are also social forms of relations between people in the process of exchange mediated by labor (Watts, 2005). Commodities are intrinsically geographic, and identifying their movement casts light on places and relations that are fixed by commodity flows such as between the city and frontier.

This chapter narrates how Laguna Lake provisions Manila with fish. It extends the resource frontier-making stories of the previous three chapters beyond the lake and into the city, establishing the urban metabolic flows and the social-economic practices that maintain urban provisioning. It presents an accounting of fish flows and identifies the constellations of actors, nodes, and relations in the chain of provisioning that connect the city and frontier. By putting commodities at the center of the narrative, I highlight the economic power structures and social relations that constitute these flows. In "socializing" urban metabolic flows, I illustrate how actors, places, and relations are brought together through everyday practices in sites of production, exchange, and consumption, using the frame of a value or commodity chain.

More than just identifying nodes and organizations of specific markets, a value or commodity chain analysis of urban provisioning identifies points of control and exercise of power among various actors, whose relations are spatially and temporally fixed through recurring, everyday social relations.[1] Value chains map distributional issues of access and benefit, within both economic and more-than-economic relations surrounding commodities. Grounding value chains and metabolic flows in place (Coe et al., 2008; Kelly, 2013) examines specific issues such as labor conditions that pervade sites in such chains (Selwyn, 2011; Taylor, 2007) and maps the institutions—the formal or informal rules of use—that structure access and control and resource trajectories (Hamilton-Hart & Stringer, 2016; Mohan, 2016; Ribot, 1998). Access, long a thematic concern for political ecological work (Cornea et al., 2016; Jepson et al., 2010; Robbins, 2011), examines place-based mechanisms of obtaining benefit and exercising power over others among various nodes in the chain. Specific mechanisms of access include rights-based access through legal and illegal means, as well as structural mechanisms, such as access to capital, markets, labor, knowledge, and authority negotiated through trust, reciprocity, and patronage (Ribot & Peluso, 2003), all of which configure the urban metabolism of food provisioning.

TRACKING THE MATERIAL FLOWS OF URBAN FISH

At the height of the Laguna Lake fishpen rush in the 1980s, estimates suggested that as much as three-quarters of freshwater fish consumed in Metro Manila was supplied by the lake. This represented a dramatic increase, since the lake had recorded a measly 0.7 percent contribution to total fish landed in

TABLE 3 Laguna Lake Fish Production and Fish Landings at Navotas Fish Port Complex, 2011

Species	Laguna Lake Production (MT)[a]			Fish Port Landing from the Lake (MT)[b]	Lake Fish Landed in the Fish Port (%)	Fish Landed in the Fish Port from the Lake (%)
	Fishpen	Fishcage	Capture Fisheries			
Milkfish	21,028	0	1,866	9,519	42	52
Bighead carp	15,182	1,746	354	12,168	70	100
Tilapia	12,284	10,548	11,558	467	1	9
Total	48,494	12,294	13,778	22,154	30	62

[a]From Philippine Statistics Authority database, 2011.
[b]From Philippine Fisheries Development Authority data, 2012, with other fish landings totaling 102,775 MT.

the Navotas fish market only years before (Alix, 1976; Guzman et al., 1974). Despite significant challenges to aquaculture productivity after 1984, the figure still stood at more than 60 percent in 2011. In the same year, Laguna Lake produced 60,788 MT of milkfish, bighead carp, and tilapia through aquaculture, four-fifths of which was supplied by large fishpen operations. Capture fisheries contributed a further 13,770 MT of these three species that are supplied in commercial volumes to the city.

Table 3 shows that a third of total Laguna Lake fish production ends up passing through the fish market at Navotas, with bighead carp and milkfish finding a more urban market than tilapia. More than half of milkfish and nearly all of the bighead carp in Navotas consumed in the city come from Laguna Lake. Milkfish and bighead carp are primarily produced by large pen operators, often within a polyculture system, while tilapia is primarily a small-scale cage aquaculture species consumed in and around the lake.

Other major suppliers of milkfish to the Navotas fish market in 2011 were the Luzon provinces of Pangasinan (37 percent), Bulacan, and Batangas (6 percent each), and suppliers of tilapia were Pampanga (63 percent) and Batangas (27 percent). Apart from Batangas, where production occurs in a freshwater lake, these other provinces produce fish in the brackish water environment of fishponds.[2] Thus, there are distinct price and quality differences associated with fish from particular places. Pangasinan milkfish are considered more desirable owing to their size and taste and would thus fetch higher prices in the urban market. Meanwhile, due to the constraints of production tied to the lake's ecological conditions, fish from Laguna Lake are usually cheaper because of their varying, nonstandardized prices depending on water conditions and are perceived as inferior in quality to fish produced elsewhere.

A key node in fish commodity flows and an important site is the urban fish market located in the Navotas Fish Port Complex (see map 4), which handled the largest volumes of fish in the Philippines and in Southeast Asia up until 2012, when it was surpassed by the tuna-landing site General Santos Fish Port in southern Philippines. Between 1980 and 2012, total volumes of fish landed in the fish port averaged 200,000 MT annually, a third of the volume of Tokyo's Tsukiji market. The fish port has been central in urban fish provisioning, as 80 percent of fish consumed in the city passes through the Navotas fish market before being distributed elsewhere in the city and throughout the island of Luzon (Celis, 1988; Sevilleja, & McCoy, 1979; Tiambeng, 1992).

From a small fish market in the 1940s, the Navotas Fish Port Complex expanded into an urban fish landing port with the postwar rise of the deep-sea industrial fishing industry in the adjacent Metro Manila cities of Navotas and Malabon. Built on reclaimed land from Manila Bay in 1976 during the martial law period, the fish port complex replaced, not without contention, the older, less formal fish market. Market transactions subsequently came under state control through the introduction of a taxation system managed by the Philippine Fisheries Development Authority, a state agency that oversees major fish ports in the country. The fish port complex's construction and changes in operational procedures allowed more efficient and timely landing and handling of perishable marine fish in the urban value chain.

The Navotas Fish Port Complex consists of five landing structures or "markets," with the longest extending 300 meters from end to end. The oldest and largest halls, markets 1 and 2, are wholesale markets, where more than two-thirds of total fish landings pass through a few large brokers. Trading begins after sunset, when temperatures have fallen, and continues through the night in time for early morning distribution to the city (Celis, 1988). The other halls house smaller daytime markets. The market halls are adjacent to the harbor, where deep-sea fishing vessels call the port to unload marine fish on the complex's bay side. The inland side, on the other hand, is where trucks from Laguna Lake and other parts of Luzon land freshwater fish.

The fish market has witnessed dramatic shifts in Philippine fisheries, the most notable being the steady rise of aquaculture and stagnation in capture fisheries. Supported by the state through subsidies and tax breaks, the

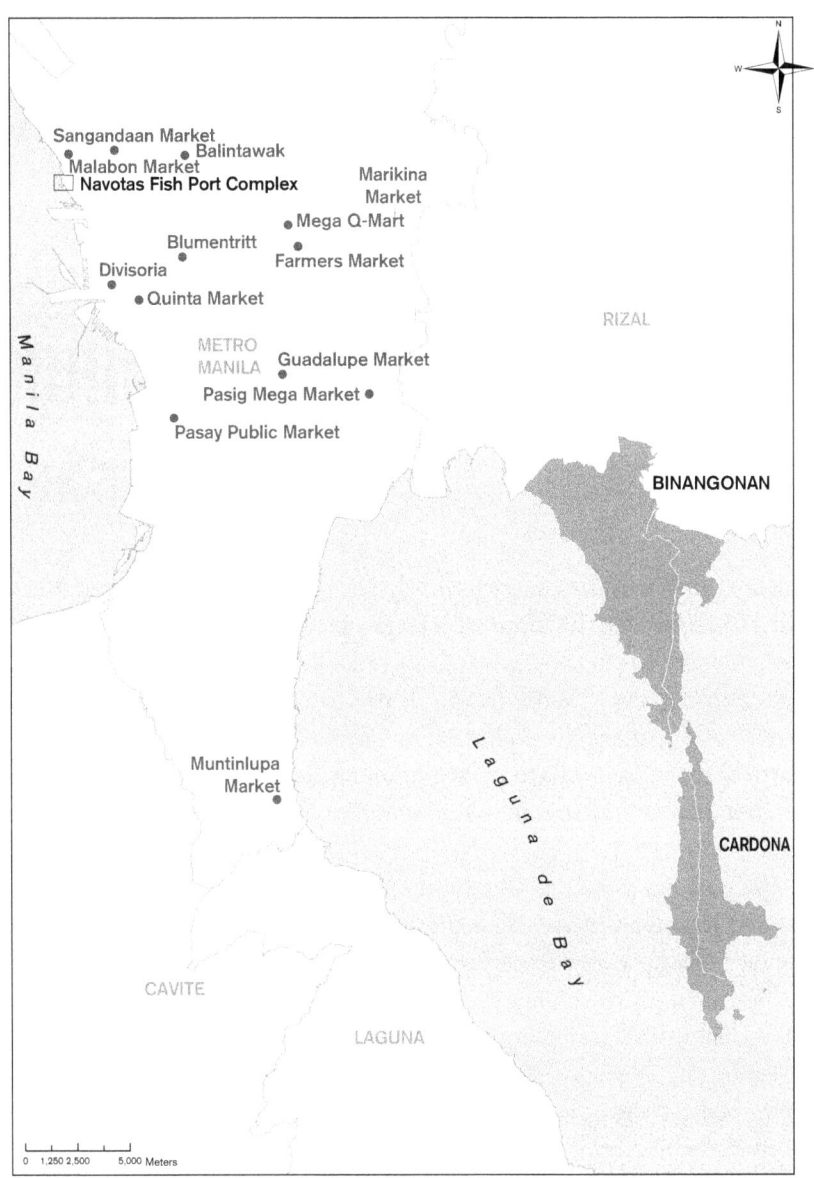

MAP 4. Location of Navotas Fish Port Complex, major wet markets in Metro Manila, and Laguna Lake fishery centers of Binangonan and Cardona. Map by Patricia Anne Delmendo.

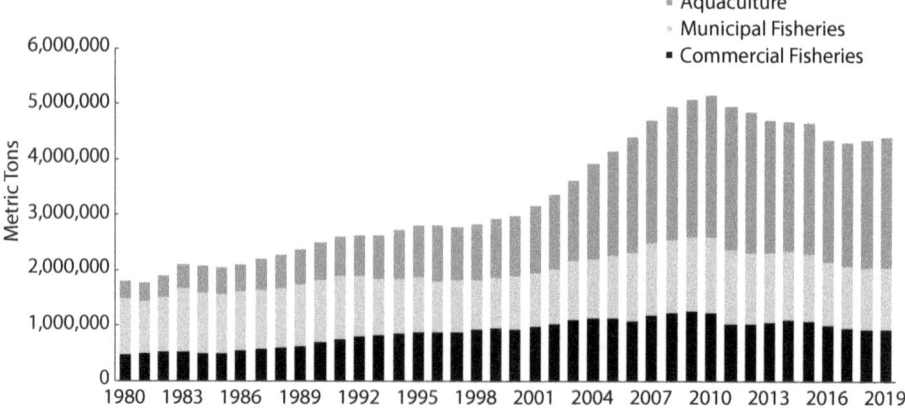

FIGURE 8. Capture fisheries (municipal and commercial) and aquaculture production in the Philippines, 1980–2019. *Source:* Philippine Statistics Authority OpenSTAT database.

country's industrial or commercial fisheries developed rapidly after World War II and became the fourteenth largest in the world in terms of fishing fleet tonnage (Green et al., 2003; Morgan & Staples, 2006; Palomares et al., 2010; Spoehr, 1984).[3] With decades of technological adoption, overcapitalization, and spatial fixes, industrial or commercial fisheries experienced a drastic slowing of productivity beginning in the early 1990s (see figure 8). This trend is in stark contrast to the rapid rise of aquaculture during the same period.

The state has promoted aquaculture as one way to meet gaps in food fish demand and supply and to relieve the pressure on wild fisheries (BFAR, 2005). Laguna Lake aquaculture has played a central role in achieving this goal since the 1970s, supplying the city with the most important freshwater fish in the urban diet. In 2012, for example, three of the top four species with the highest landings in Navotas were freshwater aquaculture fish—milkfish (second place), bighead and other carp (third), and tilapia (fourth)—compared to just one four decades earlier (milkfish, in eighth place). Farmed milkfish substituted for seasonally scarce wild marine fish in the urban markets throughout the twentieth century, particularly during the northeast monsoon months (October to February), when waters were rough (Herre & Mendoza, 1929). However, recent declines in marine fish catch have meant greater dependence on aquaculture to stabilize fish supply throughout the year.

The employment history of Tony, now retired from the fishing industry, provides insight into the changing fishing economies, the extending fish commodity chains, and the increasing corporate integration of fish supply. Tony worked for one of the largest fishing corporations first as an engineer in the fishing vessels during the company's beginnings in the 1970s. After training in Japan and working on deep-sea fishing vessels, he rose through the ranks of the company. During the height of the fishpen sprawl in the 1980s, which saw many Malabon- and Navotas-based fishing corporations investing in the highly profitable Laguna Lake aquaculture, he was put in charge of the same company's fishpen operations in the lake as it sought to expand and diversify its investments and fish supply. His task was to oversee milkfish trading in one of the municipal fish ports and manage overall pen operations. In the 1990s he was transferred to the Navotas fish market, where he worked as the company's broker-caretaker, administering the nightly whispered auctions and managing the flows of fish from both deep-sea fishing and lake aquaculture. His career path summarizes the expanding reach of large fishing corporations throughout the commodity chains of fish supplied to the city.

SOCIALIZING CHAINS FROM LAKE TO CITY

A simplified value chain of Laguna Lake fish shows flows moving from producers to traders and wholesalers via brokers to retailers and consumers (see figure 9). Two locations are important nodes in the value chain: Laguna Lake (and its sites of production) and the urban wholesale fish market in Metro Manila. As discussed in the previous chapters, actors in Laguna Lake include pen and cage operators, fish traders and seed agents, and suppliers of inputs (e.g., cage nurseries and inland hatcheries). Pens provide the urban fish market with a significant percentage of milkfish and bighead carp. These fish are sourced from fingerlings suppliers at the lake or in other areas. Agents of fingerlings buy fish from cage hatcheries and sell them with markups to pen or cage operators who are set to begin a new round of production. Some pen operators employ drag seiners and traders from the village to seine and harvest their fish, respectively. Some fish caught by capture fishers also make their way to the urban market, albeit in much smaller volumes. Trader-transporters (*digaton* and *naghahango*) bring a third of the landed lake fish to Navotas, while pen operators and vertically integrated pen corporations who land with their own brokers account for the remainder.

Brokers are the central actors in the urban fish market, mediating transactions between Laguna Lake traders or producers and wholesalers. There are

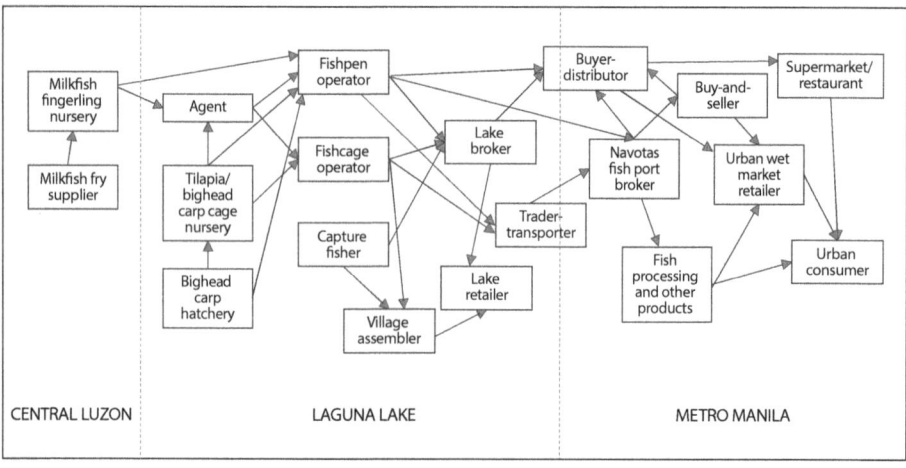

FIGURE 9. Value chain of Laguna Lake fish. *Source:* Author's fieldwork, 2012.

around seventy brokers in the fish market, with the larger ones also being owners of large deep-sea industrial fishing vessels and large pen operations in Laguna Lake. Trader-transporters typically unload their fish to small- or medium-sized brokers (also called *consignacion*), who are responsible for selling the fish to wholesale buyers. A secret auction process (*bulungan* or whispered bidding) between wholesaler and broker takes place when several buyers are interested in the fish. At the end of the trading night, unsellable fish still fit for consumption, such as poor-quality fish or low-demand bighead carp, and lake-invasive knifefish are sold to filleters, who then sell white fish fillets to processors of fish balls, a common street snack in Metro Manila. Marginal-grade and poorer-quality fish, such as those close to spoilage, are sold to fish sauce (*patis*) and fish paste (*bagoong*) makers in the Malabon-Navotas area. Wholesalers include buy-and-sellers, who have privileged access to brokers and procure large volumes of fish from them to sell to other wholesalers and retailers, and buyers-distributors, who purchase fish directly through the brokers before distributing them to retail markets, supermarkets, and restaurants.

Access and Social Relations in Laguna Lake Nodes

Overcoming high barriers to entry through access to financial capital and past pond aquaculture experience, urban fishpen operators were able to claim large sections of the lake for profitable production (see chapter 2). Through access to state officials, they were able to maintain and expand operations by circumventing regulations and through individual political connections or the collective power of their association. Capture fisheries, on the other hand, sought access to fish trapped in the pens through poaching and sabotage. Pen operators, through the labor they hired, were able to exclude fishers from their spaces with threats and the exercise of violence.

Fishpens, by virtue of the size of their production, enable large flows of fish as inputs such as seed and outputs such as harvest-ready fish that influence other producers and traders. Through access to knowledge of technology, such as contacts with an aquaculture research station staff, and start-up capital, often sourced from the cooperative, wealthier kin, urban wage work, or urban financers, several village households were able to engage in cage nursery aquaculture that supplied pens and grow-out cages with fingerlings for stocking. Involvement in this type of aquaculture transformed village relations and organization of production and enabled certain households to accumulate more wealth than they would have in capture fisheries (see chapter 2).

While some cage nursery producers are in direct contact with pen or cage operators, most rely on seed agents as intermediaries in transactions. Agents from various lakeside villages build social rather than financial capital by talking their way into creating connections (*laway ang puhunan* or "saliva is our capital"), persuading, and building trust. They actively search for or are contacted by operators of pens or cages looking for fingerlings. In turn, agents contact cage nursery producers in villages to procure enough seeds to supply the demand. After agents and nursery producers negotiate a price for the fingerlings, the former deliver the seeds to the pens with a markup and remit the earnings to the cage nursery producer in return.

This system has caused mistrust and conflicts between the two actors, especially when payments are delayed by pen producers. It has also led to depressed fingerling prices as a result of a race to the bottom when agents promise lower prices to get more fishpen clients, while some agents have also evaded responsibility for prompt or complete remuneration of nursery producers. Because there are plenty of nurseries in various villages and most transactions are paperless, agents may get away with nonpayment. Issues of trust and malfeasance pervade transactions surrounding fish, emphasizing the importance of risk perception between actors as well as access to social ties. In the case of cage nursery producers, building a base of buyers is constrained by social ties and personal knowledge of pen or cage owners. This is not easy for many village-based producers given that pen and cage buyers are situated outside the village boundaries and are often absentee owners. Those who cannot or do not build ties with seed buyers tend to reduce instances of undercounting and depressed prices.

Fishpen production also influences fluctuations in fish prices, with consequences for the income of traders, smaller producers, and even other pen operators. Gill net fisherfolk primarily catch bighead carp and milkfish, two species that are also stocked by pens. Pen harvests, often staggered over many days, tend to depress prices of fish that fishers also land in lake fish ports. Vertically integrated fishing corporations based in the urban fish market are also able to sell the fish at a cheaper price than traders who buy fish from pens and did not harvest or transport their own fish. Access to information from brokers about time and species of large fishpen landings is crucial for smaller fishpens and trader-transporters in timing harvests if they want to avoid landing when prices are low:

> This big fishpen corporation would, for example, harvest 30 tons and sell them to the buyers at whatever price agreed upon through the broker because it needs to make space for the next set of fingerlings. If you, a smaller pen operator, harvest at the same time, your 10 tons will be affected. That is why I would call and ask brokers I know how much these big corporations are unloading, and what kinds of fish. If, for example, they are unloading bighead carp, then I would try and harvest milkfish instead.[4]

> We monitor fishpen company unloading. But sometimes you think they will unload milkfish and it turns out they harvested bighead carp. So when you unload with them, you are at a disadvantage because they can unload their own fish at a much lower cost than you. Buyers will flock to them instead of you.[5]

Access to brokers to get information about prevailing market prices is especially important for trader-transporters. While freshwater fish prices remain relatively stable when compared with the seasonality of marine fish, a 2 peso change in prices due to unforeseen sudden landings by fishpen or deep-sea fishing corporations may be the difference between profit and loss.

Fishpens have retained significant rights-based access to Laguna Lake waters for production since the 1970s, historically at the expense of capture fishers. Access to authority in the form of connections with state actors, combined with strong organizational bargaining and weak state regulation, has enabled pen operators to maintain production and access to the urban market. However, the proliferation of fishpens in the lake has also created new livelihood opportunities and institutional arrangements through backward linkages, labor, and ancillary work for capture fisherfolk, including those engaged in cage nurseries, seed trade, drag seine harvest, and trade-transport. Cage nurseries benefit from pen operations through access to pen fingerling buyers, with those engaged in this livelihood becoming relatively well-off within the lake villages. Social relations of trust, reciprocity, and patronage that characterized previous arrangements pervade these emergent forms of economic connections.

Forms of preferred clientelism (*suki* relations) and patronage exist between traders and brokers or *consignacion*. In return for information on prices, traders are expected to land their fish exclusively to one broker, even if it might be more efficient for the trader to spread the haul over several brokers. Brokers are also sources of credit, especially for newer traders. Among the many fishing-based livelihoods in Laguna Lake villages, trading requires the most start-up capital. Like seed agents (and some traders use agents too), traders also need to establish contacts with pen owners for harvesting.

Market and Money in Metro Manila Nodes

Brokers have the responsibility to dispose of the fish once traders unload at the fish market. The seventy brokers in the Navotas fish market, who collectively handle more than 100,000 metric tons of fish annually, are the only actors licensed to mediate between producers or traders and buyers. Licenses or rights are renewed for at least a million pesos every five years, making brokering a high-capital venture. Most of the brokers are from Malabon and Navotas, and the largest ones have extensive linkages with traditional elite families who own deep-sea commercial fishing companies or large fishponds.

Organized into a fish broker association, brokers are oligopolistic actors able to influence others in the value chain through control of credit, information, the fish consigned to them, and the auction process (Salayo, 2000; Surtida, 2000). Eight of the largest brokers have historically controlled two-thirds of the fish market transaction volumes (Navera, 1976). Since the large brokers operate under the same corporations that own the largest deep-sea vessels and fishpens, broker control of fish flows to the urban fish market and of urban fish in general is significant. Through their employed laborers, brokers are able to access information about total fish volumes in the urban fish market and pass this knowledge on to traders and producers, including giving advice on which nights it would be best to unload fish for higher landing prices.

Brokers take a 5–6 percent commission for their intermediary work and occupy roles beyond brokering fish transactions. In many cases, they also own and control the fish consigned to them and exert considerable influence on pricing (Surtida, 2000). Well-capitalized brokers also assume a variety of risk in fish trading and derive power in the chain through their capacity to pay in cash and extend credit to suppliers and buyers (Salayo, 2000).

Traders and wholesale buyers are tied to brokers through credit relations. At the end of a night's trading, brokers pay the traders the negotiated price for the fish and transfer fish ownership to the wholesale buyers and distributors via credit payable before the next round of purchase. Wholesalers are expected to pay the cost after selling the fish and when they return to the broker for another transaction (usually the day after or the next). This creates a situation wherein trust and related relations such as *pakikisama* (fellowship or shared identity) become one of the central considerations in the transaction that sustains the economic relations (Surtida, 2000; Torres et al., 1987). Brokers only deal with established wholesalers, usually at high fish quantities. Retailers do not get access to brokers but rather to wholesale buy-and-sellers who purchase fish from brokers and sell them inside the market grounds to other wholesalers or retailers. It takes a few years before new wholesale buyers can gain the trust and credit approval of brokers.

Fish are sold to wholesale buyers through *bulungan* or whispered bidding, a secret auction system in place since the 1940s and perhaps earlier (Hinkle, 1950).[6] Late in the night, potential buyers huddle around the tubs or *banyera* (30–40 kg containers) of fish for sale and submit their bids by whispering in the ear of the bidding manager or using text messages or pieces of paper. Fish typically go to the highest bidder, but this is up to the discretion of the broker, who evaluates the creditworthiness of the buyer, which is tied to established broker-buyer relations. It is possible for traders or producers to not know the actual price of the fish sold to buyers because of the secret auction. The auction is one of the ways that brokers influence fish prices through control of knowledge. Brokers and deep-sea fishing companies also have extensive logistics and cold storage facilities both inland and offshore, and by staggering fish unloading they are also able to take advantage of fluctuating fish arrivals in the market.

Wholesale buyers-distributors, who are granted access by supermarkets to supply them regularly with fish, gain an advantage over other distributors because supermarkets buy at fixed prices and fixed volumes. However, they are often strained by supermarkets' delayed payments, putting them at risk of losing their creditworthy status with brokers who demand timely payments. Wholesalers then are put in a situation wherein they might lose both a broker as a source of fish and the supermarket as a supplier, due to late payments:

> Sometimes it takes 30 days before we get paid. If they are stingy, it takes 60 days. What if you default today, let's say a holiday, no bank is open and they issued you a check? If you don't pay the broker before the due date, you'll lose standing with the broker, and you'll find it hard to get fish again. If you failed to bring fish to the restaurants and supermarkets, they'll find other suppliers and they'll hold on to your collectibles [i.e., money owed to the wholesaler] because you broke the deal of supplying them with fish regularly.[7]

Wholesalers who distribute to other regions are typically able to control the fish markets elsewhere because of their limited numbers. Delay or nonpayment of debts is a risk that several value-chain actors constantly encounter. Also, inaccurate fish counts or volumes through manipulated weighing scales are common throughout the value chain, from seeds to retailers. Retailers pass on the costs of such practices to consumers, who even with *suki* or clientelist ties are still cheated.

LABOR IN THE FISH MARKET

Other groups that gain indirect access to the chains of provisioning need to be considered, as they constitute and shape urban metabolic relations, particularly on the consumption end and in the place where fish market exchange occurs. The Navotas Fish Port Complex is situated in Navotas, one of the densest cities in Metro Manila, in an area with high rates of poverty, unemployment, and housing insecurity, and where more than two-thirds of the city population is involved in the fishing industry, supplying the fish market with abundant labor (Navotas, 2010). Despite the abundance of fish in the market, access to benefits from fish as food and livelihood is limited for the surrounding urban poor communities, where hunger and food insecurity remains a chronic problem (FIAN-Philippines, 2009).

Residents of Navotas engage in the fish value chain as a source of income by providing labor and small, often informal, ancillary services, such as distributing ice for wholesale buyers. Scavengers or *bakaw*, which include children from nearby slums, gain access to fish by poaching or by picking up beheaded, crushed, or partially damaged fish that have fallen by the wayside

FIGURE 10. Broker laborers or *batilyo* at the Navotas fish market, 2012. Photo by author.

in the process of loading and unloading. They sell these fish in the other markets, while some edible refuse fish are sold in the shanties, becoming an accessible source of food for many poorer households. Spoiled fish are collected by scavengers to sell to fish sauce or fishmeal makers. Stealing by laborers and nonlaborers alike is common in the fish market. In response, brokers employ armed guards and deploy surveillance cameras to watch over laborers and restrict access to the fish.

Depending on their size and on fish landing volumes, brokers employ at any given time anywhere from ten to a hundred laborers, locally known as *batilyo* (see figure 10). Excluding laborers in fishing vessels that dock in the port, broker workers can be categorized into three types based on their security of tenure: regular, extra-regular, and extra. Regular workers, also called "blueboys" after the color of their uniforms, are employed with first priority, are paid relatively fixed wages, and receive health and housing benefits. Extra-regular laborers are picked by the broker's foreman on a nightly basis depending on the amount of fish to be unloaded. They are expected to remain loyal to their broker employer and receive wages based on the night's work but get no benefits. Extra *batilyos* meanwhile are only hired during times of high fish volumes. Due to their highly irregular and unseasonal employment, they often offer their labor to any broker needing workers. Access to work is stratified according to these categories, with extra laborers having the least chance of being hired during the lean season or on days with fewer landings.

These employer-worker ties are created and reinforced through patronage relations between laborers and brokers, passed on through generations of fish market work. Two fish market laborers narrated their experiences of being hired by a large broker:

> I first worked for this broker eight years ago. I knew someone who worked there so I was able to get in as an extra-regular. If you have a hook and a pair of boots, you're in, as long as you know how to pull and drag tubs of fish.[8]

> We just work for one broker because if they see us working for another, they won't like it. Maybe if you work extra for those who are far from your broker, they won't see.[9]

The abundance of available labor tied by loyalty to a large broker makes the hiring process an important aspect of access to fish market work. Dressed in the required plain white T-shirts and carrying hooked metal sticks used for dragging tubs of fish, extra-regular and regular workers sit and wait in the fish port when they are alerted that a carrier vessel is docking. The foreman then decides whom to hire for the evening based on degree of acquaintance. He then assigns chosen workers an ID with a number. The number is important as a marker of priority because lower-ranked workers are more prone to being let go early when fish volumes begin to dwindle. Laborers explained the process and how brokers get away with paying lower wages:

> If the ID numbers range from 1 to 70, it would be good if you get 40 and below because once there are few tubs left, the operators would ask higher numbers to stop working and pay them P220 instead of P300.[10]

> We worked the same amount, but received lower pay just because of the ID, just because our number is high, they would ask us to queue in another line to receive less wages.[11]

Hiring is contractual for most workers, and wages remain low, ranging from P130 to P350 per night, all below the minimum wage for the city. The working period is between six to eight hours in the middle of the night, and the risk of accidents is high, especially when laborers are told to work faster to unload all the fish in time for the trading hours and take advantage of slight fluctuations in prices. Large brokers reinforce labor discipline through guards and surveillance, meant to assess a laborer's work ethic and to prevent occasional poaching. Guards, called watchers, are also hired from the ranks of laborers and are paid extra. Due to the high value and volumes of fish and the time-sensitive nature of fluctuating fish prices, laborers experience surveillance in a variety of ways, which ensures that they work fast and efficiently and do not steal or renege on their duties:

> When they say faster, you have to go faster. They are very sensitive about time, especially when there are many fish to unload, such as thousands of tubs. You cannot slow down because they will say that you are not fit for this work and replace you.[12]

> The foreman watches over us. They also hire watchers, who are like security guards. They will hire watchers to keep an eye on us. If someone, say, went home during working time, they would tell the foreman. Or if someone is smoking while working, or took a fish.[13]

Wholesalers and smaller brokers employ fewer laborers because they require less work. Some engage in a share system of dividing part of earnings among the workers. They also employ regular workers and hire extra ones on busy nights. Because fish supply through aquaculture is more stable and less seasonal than ocean fishing, regular workers for smaller brokers dealing with farmed fish are more assured of work and income. These brokers and wholesalers are also more lenient than bigger brokers in providing laborers access to fish for take-home food. Their workers also become an informal means of access to fish as food for laborers employed by larger brokers.

The fish market creates a hazardous work environment, as *batilyos* need to navigate a slippery, noisy, and crowded workplace while carrying more than 40 kg of load for up to eight hours nightly. The low level of job security intersects with a stressful environment and high work demand to contribute to numerous accidents (Asuncion et al., 2019). Child labor is also common. Around four hundred children between ages five and seventeen are estimated to work in the fish market in 2008 doing various tasks, including *batilyo* work, scavenging, or poaching (Tolentino, 2010). Children are hired as *batilyo* laborers especially during the peak season of fish landings. Scavengers and poachers are often apprehended by guards, sometimes through violent means, whenever they are caught. Unlike the open nature of most retail and wholesale wet markets in Metro Manila, the Navotas fish market is a heavily guarded space where access is restricted and movement and behavior controlled.

Fish market laborers and fishing corporations have a long history of conflicts surrounding casualization of labor and the nature of contractual relationships. In several rulings, the Philippine Supreme Court reaffirmed the regular employment status of fish port and vessel workers, arguing that their work meets the criteria for regular employment in that they perform activities necessary to the business (RJL, 1984; Poseidon, 2006). The persistence of contractualization of fish port work perhaps points to brokers' and fishing corporations' attempts to maintain profitability despite fluctuating fish landings and to manage the particular seasonality of fish unloading (Carnaje, 2007). It is common for fishing corporations, for example, to enforce fixed-term contracts (*por viaje*) for fishing vessel laborers, who are rehired on a per trip basis. The hiring of *batilyos* is under even vaguer terms and is not based on a fixed contract but instead is subject to a foreman's decisions. The pool of labor from surrounding areas continues to be deep despite relocation of informal settlers to make way for a road-widening project in the early 2010s. Some of those relocated to the peri-urban fringe of Rizal, for example, make regular trips back to the fish port to work as *batilyo*, even when this involves nearly two hours of commuting each way.

Increasing broker control over labor is compounded by or perhaps has resulted in the weakening of a once strong trade union in the fish port. The Samahan ng Nagkakaisang Batilyo-NFL (Association of United Fishport Workers-NFL) served as the bargaining organization in dealings with the Fish Brokers Association, particularly in the 1970s and 1980s (VDA, 1993). It staged strikes when fish port work was threatened with lay-offs or casualization with the passage of the 1974 Labor Code, and with the construction of the Navotas Fish Port Complex in 1975. These events were seen as crucial moments in the urban resistance against Marcos's martial law rule (CBBRC et al., 2011; Karaos, 1993; Lumbera, 2010) and provides an early example of the agency of labor in commodity chains (Selwyn, 2009). However, laborers have noted the decreasing role of the labor organization in parallel with the increasing casualization of work:

> The fishing company let several regular workers go. They paid the older ones to leave, those who have worked for a long time. They are shying away from the responsibility of giving benefits, social security, things like that. So they hired new people, those extra ones that they can hire on a casual basis. They just give them something extra during Christmas as consolation.[14]

> Many of the laborers here are afraid to speak up because if you complain, they won't hire you anymore. If you ask for work, they won't give you one. So even those who would like to fight would just keep to themselves and endure. Whatever they give, we just accept, we just follow. We have yet to complain.[15]

Brokering remains a high capital venture limited to a few firms, owing to their exclusive rights-based access to license to operate in the fish market. Their ability to control fish flows to the city through their position at the junction of aquaculture and deep-sea fishing operations and their control of access to knowledge, technology, and credit enable them to exert influence on both upstream actors like traders-transporters and downstream ones like wholesalers. Relations of trust, reciprocity, and patronage similarly pervade the various economic relations, including such persistent practices as the secret auction or whispered bidding, and credit and suki relations. The impacts on livelihood of the fish market in nearby communities remain significant as urban slum dwellers attempt to gain access to fish market labor and illicit access to fish.

Within the urban value chain of provisioning, the largest capital investors in both production and exchange nodes deploy labor in different but parallel ways. Pen operators hire willing rural migrant labor for tasks that Laguna Lake villagers are not willing to undertake. The materiality of nature and specificity of pen aquaculture require the deployment of migrant labor throughout the year but under conditions of low wages and the requirement that laborers perform all-around pen work. On the other hand, brokers and other fish market actors employ local urban poor labor but on a casual basis. Brokers respond to the unevenness and seasonality of fish landings through the casualization of labor and often perform functions beyond brokerage as they diversify over time.

In a few significant cases, brokers and pen owners operate within one corporation in a form of vertical integration. City-based actors' driving the Laguna Lake aquaculture value chain through ownership of production and control of fish exchange shows the elements of elite capture parallel to but distinct from rural elite capture documented elsewhere (Toufique & Gregory, 2008). The dominance of urban elites in Laguna Lake and broader fish value chains is a product of complex social and historical processes, including weak or conflicting state regulation, and local place-based histories that have privileged elite access to land and capital. While it could be argued that Laguna Lake aquaculture has provided livelihoods in sites of production and exchange, as well as more affordable fish for urban consumers, distribution of benefits continues to be highly uneven and is reinforced by persistent institutional arrangements and everyday practices of exclusion and restricting access.

CONCLUSION

Mapping the series of relations as commodities exchange hands in urban fish provisioning illustrates how urban frontier making extends back to the city through production and maintenance of resource flows. Urban metabolism in Manila and Laguna Lake is constituted by social relations that are mobile but also fixed in place through economic transactions and social ties in everyday livelihood making and structured by enduring political economies of urban fish production and exchange. Practices of gaining, maintaining, or excluding access to and control of urban fish commodities emphasize how metabolic resource flows are thoroughly socialized and embedded in relations of power.

Frontier making is not just about transformation of the frontier landscape but also about how established and emerging city-frontier relations shape places along the chains of resource flows. Sustaining the city with fish involves more than just stories of material flows and volumes; it also includes narratives of access to and control over commodity flows in sites of production and exchange. Nature is not merely a material substratum transformed in sites of production to deliver flows but also emerges as an active site of coproduction through labor and practical activity along the commodity and value chain. In the next chapter I pair these urban fish supply stories with narratives of how city dwellers consume fish and attempt to negotiate with the ecological contradictions in the chains of urban provisioning.

FIVE

Biographies of Fish for the City

THE HUMBLE BIGHEAD CARP is an unlikely window into the ecological contradictions of frontier urbanism in Manila and Laguna Lake. Its disproportionately large head and mouth betray its hardiness and adaptability, which have made it an unintended staple of urban fish provisioning and an unintended solution to particular urban metabolic problems. Tales of bighead carp and other ordinary farmed fish consumed in much of the Global South often do not get told in the same ways that magisterial histories of the Atlantic cod fisheries' collapse or of the growth of Pacific Northwest salmon fisheries do. Yet the bighead carp is not an unimportant fish. In 2016 it registered the fifth highest production among the world's farmed fish, trailing only a few other Asian carp and the Nile tilapia. It outnumbered boom crops for export such as the Atlantic salmon and pangasius, having been introduced in seventy-two countries throughout the twentieth century. But as a neglected species (Belton & Bush, 2013), how bighead carp have traveled and been encountered in various places is no less fascinating. In its native China, the fish is a valued staple of restaurants and local cuisine, often commercially farmed with other carp and fish (Kolar et al., 2005). In the United States, introduced with a nonfood purpose of improving water quality since the 1970s, the fish is seen as a pest that needs to be contained, a threat on the brink of invading the commercially important Great Lakes fisheries (Cidell, 2018; Cooke, 2016).

In Laguna Lake, fishpens have cultivated the fish for nearly half a century. While it does not fetch the highest value for producers when harvested, it is a choice crop, especially during years of suboptimal water conditions. In 2014 the lake produced more than 15,000 metric tons of bighead carp, nearly all of which passed through Metro Manila, where it circulates as the cheapest fresh fish. Despite these substantial volumes, the bighead carp is largely invisible in many of Manila's wet markets and in the wider rich Filipino cultural imagination of fish. That the fish is also known by many names—*mamali*, Taiwan, Imelda, and *maya-mayang tabang*—presents a way to explore how the fish is encountered in everyday life and how this embodies Laguna Lake's particular urban metabolism.

This chapter presents a biography of the bighead carp to illustrate urban metabolic connections between city and lake. It builds on urban provisioning accounts in chapter 4 by demonstrating the ecological contradictions of urban resource frontier making embodied in the production and consumption of a particular type of fish. Borrowing the term *biography* from Kopytoff (1986), it reveals the social lives of food commodities as they travel within states and across sites of commodification. Bighead carp acquires different meanings and is materially transformed in multiple ways in sites of production, exchange, and consumption. Tracking bighead carp's circulation shows that the modern vision of producing more fish confronts ecological problems rooted in the urbanizing material characteristics of the lake.[1] As producers are able to find ways to circumvent these problems to produce more fish, they encounter further contradictions in the city, where the glut of unfamiliar and undesirable fish requires transformation by city dwellers to make the fish consumable. This story uncovers both aquaculture's limitations as a substitute for wild fisheries and the diverse material practices of making do that constitute urban resource flows from the lake. It also reflects questions of social reproduction and class relations mediated by socionatural transformations and embedded in city-frontier relations.

Unlike other vital urban flows such as water, food plays a distinctive role in social reproduction, defined as the varied practices and processes by which labor power and the means of production are reproduced in relation to capitalism (Bakker & Gill, 2003; Katz, 2001; Mitchell et al., 2004). The circulation of food as the material basis of social reproduction shapes urban metabolic configurations that are produced and that reinforce particular capitalist relations, as city dwellers rely on commodified food from elsewhere.[2] But the movement of food flows from frontier to city requires work as they are transformed at various nodes in the chain, often by acts of improvisation that contribute to sustain metabolic relations. The story of the ordinary bighead carp produced in Laguna Lake and consumed in the city presents an entry point to understand the contradictions of urban provisioning and their implications for these socioecological questions.

"IMELDA" AND BIGHEAD CARP IN THE LAKE

Bighead carp arrived in Laguna Lake in the 1970s by way of Taiwan, having been introduced into the country by the Fisheries Bureau in 1966 (Baluyut, 1989). It has been stocked in other lakes and reservoirs to augment fish production, mainly during the Ferdinand Marcos presidency, and has thus acquired its colloquial name "Imelda" after the former First Lady. Its other local names, *mamali* (threadfin salmon) and *maya-mayang tabang* (freshwater red snapper) may have pertained to its physical resemblance to other types of marine fish (see figure 11). Unlike the common carp, which has been present in the lake since its introduction during the American colonial period, the bighead carp can be farmed and its reproduction controlled, enabling high volumes of production. This greater control in production has been facilitated by the development of techniques of induced spawning and fry rearing, which in turn has created a specialized industry in the lake.[3]

A key figure in this industry is Ramon, one of the pioneers of bighead carp culture in the country. In his lakeside house, he was quick to show me his backyard hatchery, where the fry are spawned and reared to fingerling size. Inside, he gave me a tour of the important equipment he had slowly accumulated over the years—oxygen tanks, incubators, recirculating tanks—emphasizing along the way techniques he employs to keep the water properly aerated and avoid mass mortality. He spoke with pride about his bighead carp experiential knowledge and expertise, honed by over three decades of trial and error and painstaking note taking. Like many others in his lake village, Ramon took up tilapia fishcage culture in the 1980s, but he was one of the few who switched to bighead carp fry production years later. Barriers to entry into this more knowledge- and capital-intensive industry were high. Only two others at the lake—and perhaps in the country—operate a bighead carp hatchery on a similar scale.

Scientists and television reporters alike, he mentioned while thumbing through his stack of thick notebooks, would come to him to learn about bighead carp reproduction despite his lack of formal fisheries education. He would rehash the same lines, reiterating the importance of bighead carp in providing cheap and abundant fish for food security. For him, operating a bighead carp hatchery has been highly profitable, particularly with the tenfold increase in bighead carp production from the 1990s to the 2000s. Large-scale fishpen operators and small-scale fishcage nurseries at the lake (and elsewhere) would order bighead carp fry in the millions. He described

FIGURE 11. Transporting bighead carp, 2012. Photo by author.

why, despite fetching low farmgate prices, the fish was especially important to the survival of Laguna Lake aquaculture:

> Some fishpen operators tell me, "You know without you, many fishpens would have ceased operating, many of us would have left Laguna Lake. We cannot rely on milkfish and tilapia because when water conditions deteriorate, milkfish won't grow, sometimes they would even lose weight." But bighead carp does not mind water conditions. It grows well whether the water is turbid or clear, cold or warm; it is able to grow. That is the advantage of bighead carp. It is an all-weather type. So, many fishpen producers who stock

milkfish also stock bighead carp. Even the biggest fishing corporations there stock bighead carp.[4]

For fishpens, stocking bighead carp is a way to manage the risks of lake materiality where fluctuating water conditions can prolong the production time of fish. As detailed in chapters 1 and 3, Laguna Lake's fish productivity depends heavily on its water quality, transformed by various urban activities and infrastructural interventions. The lake is highly eutrophic due to the abundance of nutrients, a property that has enabled aquaculture to produce fish at a comparatively lower cost than other places. Reliance on natural plankton for feeds, however, has limited the ability of producers to intensify and exposed them to recurring risks of poor water quality, including the threat of "bad water," heavy metal contamination, and prolonged growing seasons in the absence of saltwater intrusion from Manila Bay. Fish produced in the lake therefore tend to be of varying rather than uniform sizes, as a result of these multiple ecological factors. They also acquire a seasonal earthy-muddy taste due to annual *Microcystis* algae blooms and are perceived to be of inferior quality to fish raised with feeds elsewhere.

The bighead carp's tolerance for less-productive water conditions—less saline and therefore turbid and fewer nutrients—enables it to replace higher-value and more-preferred milkfish at certain times. While milkfish tend to be sensitive to water quality and nutrient availability, taking more than a year to reach market size in some instances, bighead carp are able to grow in any condition. This difference was most pronounced in 2001, when bighead carp production doubled from the previous year to offset a significant decline in milkfish production in fishpens. Since then, the two species have been cultured in tandem or in succession (see figure 12). Fugitive bighead carp from pens are also caught by capture fishers, who often sell them in the fish port as they are not usually consumed for subsistence in the lake, unlike milkfish and tilapia.

In many ways, bighead carp have provided a fix for the long production time of milkfish in fishpens that results from poorer water conditions in the lake. Pens can circumvent the problem of long gestation times of higher-priced crops like milkfish by cultivating the hardy bighead carp instead. However, this is not a perfect solution, as bighead carp fetch around one-third the price of milkfish for the same weight. Furthermore, despite a lot more fish being produced in the lake, consumption in the city has not been as widespread. I turn later to why this is the case and point to processes that make the fish more consumable in the city, distancing its freshwater lake nature and

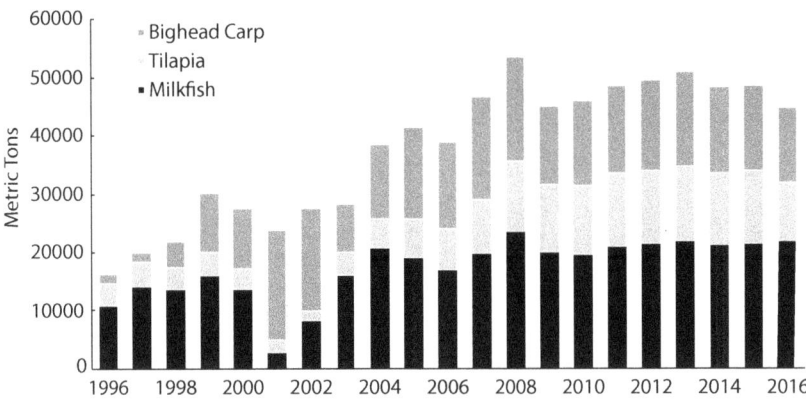

FIGURE 12. Fishpen production by species in Laguna Lake, 1996–2016. *Source:* Philippine Statistics Authority OpenSTAT database.

entangling with forms and qualities more desirable and familiar to consumers. But first I need to situate the context of cheap farmed fish in Metro Manila in the crisis of marine fisheries and the problem of feeding the city.

CHEAP FISH IN THE CITY

Fish and seafood are an important source of sustenance in archipelagic Philippines. Fish comprised more than half the average daily protein intake in 2008 or 36 kg per person annually, which is more than twice the global average. Historically, urban residents derived as much as 80 percent of their animal protein intake from fish and seafood and preferred them for everyday meals over meat (Doeppers, 2016). The proportion of fish consumption relative to other animal protein sources has decreased, however, for several reasons, such as diets changing in favor of cheaper chicken and pork and the sustained price increase in traditional marine wild fish. More expensive marine fish is the result of stagnant production due to full or excess exploitation of commercial and municipal fisheries, a crisis that increased aquaculture production seeks to address (BFAR, 2005; Dey & Ahmed, 2005). In urban Manila, aquaculture's emergence is best exemplified by three farmed fish species—milkfish, tilapia, and bighead carp—replacing marine species as the most unloaded fish in the urban wholesale market and, consequently, the species that fetched the lowest prices.

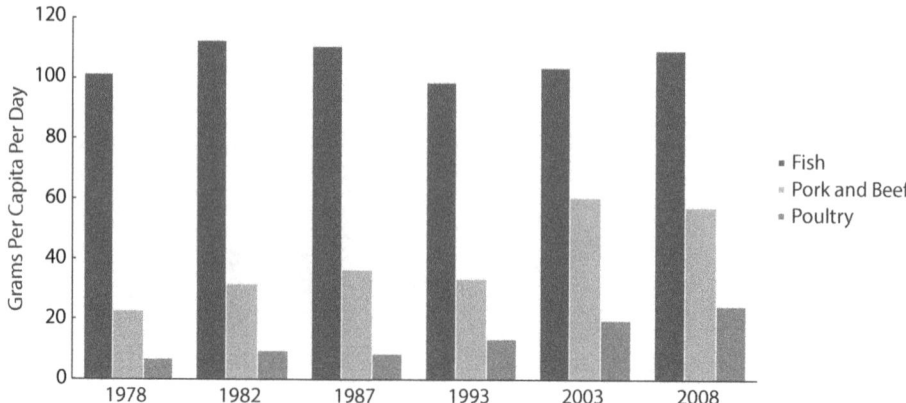

FIGURE 13. Animal protein intake in the Philippines, 1978–2008. *Source:* Philippine Statistics Authority OpenSTAT database.

Consumption of fish, considered the "poor people's protein," follows along socioeconomic class lines (Garcia et al., 2005; Yosef, 2009). In Metro Manila (as in other parts of the country) the lower and extremely lower income groups that make up two-thirds of the urban population proportionally consume more fish than other meat. The type of fish consumed matters as well (see figure 13). Round scad or *galunggong* (*Decapterus* sp.), a marine wild fish, has long been considered a symbol and indicator of poverty. With escalating prices of this fish, however, other more common and previously "middle-class fish," such as tilapia and milkfish, have become cheaper. In a symbolic move in 2003, President Gloria Macapagal-Arroyo proclaimed tilapia, a farmed fish, the new national staple fish in place of round scad (Yosef, 2009).[5] In 2012, round scad's prevailing retail prices equaled or exceeded those of milkfish, tilapia, bighead carp, and even meat like chicken.

Significant volumes of fish pass through Metro Manila, and Laguna Lake serves an important role in mediating these flows. Owing primarily to the materiality of production—the presence of natural plankton and absence of feeding—the lake produces the cheapest of the fish species sold in the city that are accessible to the more than eight million who belong to the lower and extremely lower income groups (see figure 14).[6] The urban poor often access fresh fish through wet markets (*palengke*), neighborhood markets (*talipapa*), and peddlers (*naglalako*), especially those households without refrigeration.

Large-scale Laguna Lake fishpen aquaculture not only produces lower-cost fish but also supplies the urban market with varying sizes of fish as

FIGURE 14. Animal protein intake by income class in Metro Manila, 2008–2009. *Source:* Philippine Statistics Authority OpenSTAT database. Note: Fish include round scad, milkfish, and tilapia.

a result of the fluctuating water conditions. In times of poor conditions, producers are forced to harvest fish even when they are smaller than regular market size to reduce nonproduction time and hasten the turnover cycle. These small fish find a market among poorer urban households that are able to purchase more pieces of fish per kilo. In 2012, considered by pen and cage producers a poor year in tilapia production, small tilapia (less than 100 g) were abundant in the urban market, mostly coming from Laguna Lake. A neighborhood market retailer noted the popularity and profitability of selling smaller fish in poorer communities. Households could purchase more pieces of fish for the same weight, which were distributed individually among members during meals.

What kinds of food eaten by Manila's population—in terms of type, size, cost, taste, and cultural attachment—matter in food provisioning and the consequent environmental transformation of urban resource frontiers? The case of the variously sized tilapia reflects the role of Laguna Lake as a provisioning space that is important in supplying cheap and accessible fish for much of the city's population. The story of the bighead carp is in many ways similar—cheap fish produced in vast quantities in Laguna Lake bound for the city—but it encounters unique problems of its own.

"FRESHWATER RED SNAPPER" AND BIGHEAD CARP IN THE CITY

Its bland but fishy taste and the presence of many bones limited the initial urban acceptability of bighead carp in the 1980s (Baluyut, 1989). Since then, various practices of value chain actors in the spheres of exchange and consumption have attempted to make it more palatable and desirable as food. In what follows, I describe examples of how such practices are performed in the spaces of urban retail (wet markets), cooking (kitchens), and fish processing (fish ball manufacturers).

Chopping

Bighead carp is the only fish—freshwater or marine—commonly sold in the wet markets of Metro Manila that is displayed chopped, with the fish body sliced into thick cuts and the head priced separately. Several stalls that sell tilapia and milkfish mention the place from where they originated as a way to indicate quality—often Batangas and Pangasinan, respectively. However, the stalls that display bighead carp only show prices, excluding origin. This is unsurprising given that almost all of the bighead carp sold in Metro Manila come from Laguna Lake, a place that is often associated with inferior quality: blander and sometimes with an earthy-muddy fish taste. When selling milkfish and tilapia from the lake, for example, retailers do not specify where they are from. Bighead carp too are often both nameless and placeless in wet market fish exchange.

Chopping and minimal labeling are retail strategies not only to distance the fish from its lake origins but also to make it resemble red snapper, *Lutjanus sebae* or *maya-maya*, a more desirable and expensive marine fish. Labels in stalls that name the fish include *maya-may*a (red snapper) or *maya-mayang tabang* (freshwater red snapper), a new term that originated in the wet markets of Manila only within the last two decades. Both the marine and the "freshwater" red snapper have bright pinkish flesh and a similar shape and size. The notable difference is the bighead carp's disproportionately large head, which is often sold separate from the fish body at a lower price. A fish market worker noted this practice and how it is easy to confuse the fish with something else, the "real" red snapper:

> In the wet markets, people are not able to distinguish the fish. When sliced and filleted, its flesh resembles that of a red snapper. That is why when bighead

carp is sold in the market, the head is displayed face down, not up, because one is not able to note the difference between bighead carp and red snapper. If you are not familiar with the fish, you would mistake it for the real thing.[7]

Cooking

Retailers encourage consumption through various culinary suggestions. These are sometimes indicated on the labels or are mentioned by retailers in the process of attracting customers. Unlike other fish, bighead carp requires a proper way of preparation to wash out the strong fishy taste and particular cooking methods suited to the blander flesh and larger head. This extra required work and access to knowledge have deterred some from consuming the fish despite its affordability. Making the fish more palatable through additional steps in cooking shifts the task of metabolizing the fish to the spaces of homes, where work is often performed by women or household helpers who need knowledge of proper ways of cooking to be able to make the fish taste better. The process involves washing the fish with salt before soaking it in ginger and adding *pandan* leaves for fragrance. This knowledge and work is one factor that determines those who will continue to consume the fish. A retailer and a consumer explained the work necessary prior to fish consumption:

> Consumers here prefer tilapia and milkfish because only those who know how to cook *maya- maya* buy it. It used to be hard to sell it here [a neighborhood with a large informal settlement population]; everyone ignored it. So you have to instruct consumers how to cook the fish so that they come back. Eventually, more people brought *maya-maya* when they learned about specific ways to cook it. *Maya-maya* has a fishy taste if cooked improperly but it tastes good when prepared well.[8]

> We tried *maya-maya* before. But I found the fishy taste off-putting. And I had to do several things before my family could eat it. I don't know how to properly cook it to get rid of the fishy taste. I just fry the fish or grill it or cook it in vinegar.[9]

Parts of the fish are cooked in different ways, with the head used in *sinigang* (tamarind-flavored sour and savory soup) and the more versatile flesh for frying and in *escabeche* (fried fish topped with sweet and sour sauce), among other uses. In the example of *sinigang sa miso* (sour soup with a Japanese seasoning), a popular dish served at home as well as in restaurants and sidewalk eateries, bighead carp head serves as a cheaper substitute for the more expensive yellowfin tuna. In this sense, not only are retailers passing off bighead carp as an altogether different marine fish (and thereby attaching to it its desirability and premium price), but consumers are using the fish as a cheaper substitute for other kinds of marine fish:

> We got used to cooking *maya-maya* in *sinigang sa* miso. We only buy and use the head for cooking because we find the flesh not exciting to eat. The head though has different textures, some of it soft, and then you have the eyes.... When we learned about Imelda, I would go back to where I first bought it and request more. Price is a factor too. Yellowfin tuna head is more expensive. One gets more *maya-maya* per kilo.[10]

Bighead carp consumption in Metro Manila is associated with class, but in a different way from other places like Mexico, where consumption of introduced carp in place of native species is tied to identities of being modern in contrast to being poor and indigenous (Tapia & Zambrano, 2003). Bighead carp in Metro Manila is popular in markets that cater to the urban poor, and this increased consumption partly accounts for the fourfold rise in wholesale prices.

Mel, a fish retailer in Sauyo, Novaliches, an area in northern Metro Manila where a quarter of the fifteen thousand households live in informal settlements, notes that while bighead carp is not as popular as milkfish and tilapia, he is able to earn more from selling it because more people can afford to buy the fish. He is also able to sell it with a comfortable markup. While a kilo of bighead carp costs P33 in the wholesale market, its flesh retails at P70–100 and its head at P40–70. The markup gap is partly explained by the low fillet yield of the fish given its large head, and also by the premium attached to its resemblance to a higher-priced fish.

The deceptive retail practices, the amount of work to make the fish taste good at home, and the class connotations associated with its consumption are three reasons that deter some from consuming the fish. A middle-class fish consumer's quote summarizes the problematic process of bighead carp substitution:

> I tried it once or twice. No offense but I am not too poor to buy this fish; it is really cheap. I also feel bad for my household help who prepares it in our house. It really takes a lot of work. Third, I don't like this feeling that I am being deceived, even if the retailer had told me it is not the real *maya-maya*.[11]

Processing

Prior to gaining greater urban consumer acceptability as food fish in the 2000s, bighead carp produced in Laguna Lake was primarily used for food processing as extenders and fish fillet substitutes (Baluyut, 1989). State-supported research also developed it as one of the first freshwater fish for surimi production (Fernandez et al., 1998). Its most common use in processing, however, is in the production of fish balls, a common street food sold by hawkers in wooden carts, where the fish balls are deep-fried, skewered, and dipped in sauces. Fish balls as a street food are a cheap, quick, and accessible snack for the urban working class and students, usually consumed in the afternoons after work or school (Tinker, 1997). They are also popular snacks in urban poor neighborhoods, where they are sold fried in neighborhood variety (*sari-sari*) stores.

Fish balls, Cantonese in origin, are made of white fish, flour, and flavorings. The proportions of the ingredients determine its quality. To keep fish balls cheap, manufacturers use more flour than fish and use the cheapest available white fish of the lowest quality available at the urban wholesale fish market. In some areas of the Philippines, traces of pig and chicken meat, which are now cheaper than most fish and seafood, have been documented in fish balls and squid balls as manufacturers attempt to keep costs even lower (Sarmiento et al., 2018).

Prior to the manufacturers' use of bighead carp in the 2000s, the low-priced marine fish *kalaso* or lizard fish (*Saurida elongata*) was traditionally used as the white fish ingredient for fish balls manufactured in Metro Manila. The fish's increasing price, resulting from decreased wild stocks and increasing costs of commercial capture fisheries operations—even for a marine trash fish like *kalaso*—has made freshwater bighead carp more attractive. The result is a change in the taste, volume, and texture of fish balls. Balls that used to be round and fluffy when deep-fried are now flatter and do not taste as good, something that a number of consumers have clearly observed.

The invasion of Laguna Lake by knifefish, with populations peaking between 2012 and 2016, also created opportunities for manufacturers to utilize an even cheaper source of white fish (chapter 3). Knifefish are caught in vast quantities in the lake as they are considered carnivorous pests by both fishpens and capture fishers alike. Like bighead carp, consumption of the fish at the lake remained very limited, and the invasive fish has eventually found a market among fish ball manufacturers, given its similarities to bighead carp as white fish. It fetches even lower prices due to its abundance as a result of state retrieval programs in the lake.

The example of fish balls shows how processing is one practice to enable the urban consumption of bighead carp, primarily by the urban poor, by transforming the fish into an unidentifiable form, distancing it from its freshwater nature, and entangling it with other ingredients. It is also an example of the freshwater aquaculture substitution of marine fish, one that had corresponding material effects in the taste and form of the fish balls.

CHEAP FOOD AND ECOLOGICAL CONTRADICTIONS IN URBAN FLOWS

At the height of record inflation in late 2018, the secretary of the Department of Agriculture announced that it would be importing round scad from China to stabilize increasing fish prices. This was met with public dismay that the traditionally rich marine waters of archipelagic Philippines could not supply enough fish for its people, particularly its growing urban population dependent on food produced elsewhere. Discourses about the place of aquaculture in addressing this issue reemerged.

The crisis in marine fisheries and its implications for meeting food security have long been framed as a problem that aquaculture can fix (Dey & Ahmed, 2005; FAO, 2006). In many ways, the growth in the production of cheap farmed fish in the Philippines has addressed parts of this problem. But as documented in other contexts, these efforts have created new contradictions, including increased volumes of cheaper farmed fish that undermine price premiums of wild fish, new forms of environmental degradation different from that of capture fisheries, and materially unhealthful fish (Mansfield, 2011). In the spaces of fish consumption in cities such as Manila, the substitution of farmed fish for wild fish is neither simple nor complete. The example of bighead carp is illustrative.

Introduced to improve aquaculture production in the lake for urban consumption, bighead carp did not become a common fixture on the plates of city dwellers, even those of the poorest, until the 2000s. This was due to consumer unfamiliarity with the fish and to the bony, bland, and fishy character of its flesh, partly associated with its freshwater and lake origin. In this sense, bighead carp encountered "frictions" in its flows to the city that actors in the exchange and consumption nodes attempted to overcome through practices of distancing and entanglement.

Distancing and entanglement are crucial in the process of smoothing frictions in metabolic flows of bighead carp. In wet markets, through filleting and labeling with a new name, retailers distance the fish from its freshwater lake origin and entangle it with red snapper, a higher-priced marine fish. Whether intentional or accidental, this practice relies on misleading consumers in an attempt to make the fish sellable. In kitchens, culinary suggestions made by retailers and others enable those who purchase fish to consume it more palatably. In this sense, cooking—the process of bringing the "natural" outside world to the internal domestic world where socionatures are ultimately metabolized (Chevalier, 1998)—transforms the fish into food through the distancing of the taste and character of its flesh and by entangling it with other, more familiar and entrenched ways of consuming fish. But this process involves additional steps in the preparation and cooking of the fish. This extra work aligns with household division of labor, which in some cases involves a gendered and classed transfer of the burden of metabolizing the fish in domestic spaces. In fish processing, the white meat character of the flesh allows it to be stripped of its freshwater lake origin and become entangled with generic white fish for fish ball processing. The result is a cheap but slightly different final product, a flatter and blander fish ball.

The practices of distancing and entanglement reflect the incomplete and messy process of replacement of wild marine fish with farmed freshwater fish and of urban fish resource frontier making, where extra work is required in the everyday acts of selecting, cooking, and consuming fish. The freshwater origins of the fish persist despite attempts to make it resemble something else. The blander taste of the meat and the slight change in form of fish balls are material examples, as well as the seasonal earthy-muddy taste characteristic of Laguna Lake *Microcystis* algae blooms that occasionally appears with the fish. Less tangible but more worrying are findings suggesting that certain fish in the lake bioaccumulate heavy metals from surrounding industrial activities and hormones from urban organic wastes, with corresponding public health impacts on regular fish consumers, who are often lake-area dwellers and the urban poor (Molina, 2012; Molina et al., 2011; Paraso & Capitan, 2012). These examples lend support to the claim that aquaculture produces a materially different fish (Mansfield, 2011) that embodies a different set of contradictions tied to its production and distinct from the fisheries it seeks to replace or substitute for.

Beyond the farmed-wild relation, Laguna Lake fish illustrate that food practices create the work of maintaining urban food provisioning. Encountering the ecological contradictions rooted in the materiality of production extends across sites of production and consumption. While fishpen producers are able to switch to other species such as the bighead carp, consumers make do with what is available in the wet markets, necessarily transforming them to make them edible. This shows that urban food flows are neither static nor stable streams of commodities that smoothly travel between production and consumption. Rather, similar to arguments made in infrastructure studies, vital urban flows are constituted by the work of maintenance and improvisation that sustains their continued functioning (Graham & Thrift, 2007; Howe et al., 2016; McFarlane et al., 2014). Everyday practices of circulating fish commodities pass through various hands and are encountered, reworked, and transformed in multiple ways. Urban metabolism is therefore always actively constituted by people's work and labor in attempts by city dwellers to make do, thereby mediating nature-society and city-frontier relations (Ekers & Loftus, 2013; Smith, 2008).

Bighead carp (and cheap fish) flows to the city from Laguna Lake provide a low-cost means of continuing to secure the social reproduction of urban labor power. The city consumes surpluses produced in the countryside through the least cost and effort for appropriating "nature's free gifts" in frontier zones (Moore, 2011), allowing the production of cheap food that reduces the cost of reproducing urban labor power. In this sense, the fish cheaply produced from Laguna Lake can be regarded as an attempt to provide a stream of wage food for the urban working class in a growing Global South city, which has expanded in size due to migration of rural peoples, many of whom were displaced from their lands through parallel processes of primitive accumulation and proletarianization in the countryside (Ofreneo, 1980). These flows of cheap fish continue to flood the city, but with the opening up to fish imports might be further reshaped by global flows of cheap food from elsewhere.

CONCLUSION

By focusing attention on the undesirable, invisible, and often ignored bighead carp and its urban circulation, we see the ongoing shifts in urban metabolic flows that produce new human-environment relations, new fish bodies, and new natures. Practices of making do and improvisation rework the nature of this urban metabolism as much as the changing ecological relations in sites of production and consumption of food commodities.

Bighead carp, consumed primarily by the urban poor, serves a dual role: one as a fix to the pen production problems associated with relying on plankton and another as a low-priced commodity that reduces the cost of social reproduction of labor. A significant cost of provisioning for the city is therefore borne by Laguna Lake as the frontier transformed to deliver cheap fish for the city and city consumers, who undertake the everyday work of transforming commodity flows. In a parallel way, the production of uneven benefits and burdens of frontier-making similarly surrounds stormwater flows, the spatiotemporal production of which I trace in the next chapter.

SIX

Infrastructures of Risk

A FEW MONTHS AFTER one of the worst floods in its history struck Metro Manila in 2012, I paid a visit to the operations center of a flood control infrastructure that managed stormwater flows in the city.[1] Anna, a state engineer, showed me a map of the stations that recorded water levels and pointed to the channel capacities of different streams within the hydrological network of Metro Manila. She explained the details of the complex hydrological design and the need for synchronizing flood control operations through infrastructural control from this particular site and a few others in the city. The Lower Marikina River, which cuts through a valley in the eastern half of Metro Manila, has a channel capacity of about 3,300 m^3/s, bringing water from the Sierra Madre Mountain Range through the city. A flood management dilemma emerges for state engineers when the Marikina meets with the Pasig River, which could only accommodate a fifth of this total streamflow. During periods of intense rainfall, therefore, remaining excess flows of about 2,700 m^3/s needed to be channeled away from the Pasig River to spare Manila's populated sections from disastrous flooding. The operations center we were in and the nearby weir that connects to Laguna Lake through a floodway are part of a larger network of flood control infrastructure built precisely to oversee this diversion. The idea of stormflow diversion as designed, she explained to me, was "so people downstream won't suffer too much from floods, because economically, if you plot it, the population is concentrated there."[2]

For the flood control infrastructure to work and keep the city dry, stormwater flows must be diverted away from the urban core and into Laguna Lake to its southeast, turning the large body of water into a temporary stormwater reservoir. This happened in 2012 when engineers had to close the floodgates to save Manila from the possibility of worse flash flooding after monsoon rains dumped near-record rainfall. During the time of this particular visit to the flood control operations center, Laguna Lake had already been flooded for ten weeks, and water levels that rose to double the lake's volume would not return to average conditions until December, some two months later.

Images of disrupted lives greeted me several kilometers southeast in the Laguna Lake village of Navotas, where up to a foot of lake water had invaded most homes weeks earlier. *Lampitaws* or outriggers plied the streets instead of the usual motorcycles, ferrying villagers who otherwise would have had to carefully wade through the near knee-high floodwater. Over a meal of *adobong kanduli*, Julie, my host, recounted how the floodwaters had risen slowly in her house a few weeks earlier and had stayed high since then. The turbid waters had cleared enough by then such that I could see a small school of tilapia fry swimming in her living room. She mentioned that she had lost her appetite from seeing her house flooded as she and her family tried to maintain a semblance of normal everyday life and resume making a living off the lake.

From Julie's lakeside house in Talim Island, I noticed submerged aquaculture structures and fish corrals in the distance, some of which had been damaged by clumps of water hyacinths that had clustered around their perimeters. Fishers continued to head out to sea later in the afternoon to catch milkfish that had escaped from fishpens, which initially were abundant as soon as the monsoon rains had stopped. But with the escaped stock waning, they had little to rely on. The impact on their livelihood of the lake swelling caused by stormwater from elsewhere reminded lake dwellers of the urban infrastructural control of Laguna Lake. Over the succeeding weeks, fisherfolk groups routinely called for the opening of floodgates downstream, which they believed were responsible for the extended lake swelling, as these prevented the outflow of water to Manila Bay.

Like the historical frontier-making interventions described in chapter 1, these contemporary encounters revealed an aspect of the imaginaries of an urban frontier and the ecological work such spaces are expected to perform. Socioecological risk takes place as a result: risk is spatialized and territorialized through material flows from the city and the construction of Laguna Lake as a particular type of space in relation to the city. Urban flood disasters have erupted with increasing frequency, especially between the years 2009 and 2013. But the production of urban flood risk has deeper historical roots, stretching back decades to colonial Philippines, and intertwines the city and the lake through flood control infrastructure and flows.

Presenting parallels to the frontier-making narratives of fish provisioning in the previous chapters, this chapter narrates the construction of imaginaries of the lake as a noncity sink for stormwater designed to solve persistent urban drainage problems by transferring risk elsewhere. It illustrates the intersecting multiplicity of urban metabolic flows produced in frontier urbanism. As Anna, the city engineer, had shown me, understanding flood risk in Manila must be situated within the context of the hydrological spaces and engineering interventions in the city and beyond, as well as the bureaucratic work and infrastructural politics associated with which sites would get flooded, by how much, and why. The question of urban socioecological risk, like urban provisioning, has been understood in multiple ways as cities expand and enroll metabolic flows with increasing distance. This suggests that the centrality of constructions of risk mobilize notions of riskiness and risky landscapes that people encounter and contest at a material-ideological level in cities and their frontiers (Mustafa, 2005). Risk not only emerges as a property of space or landscape but is produced through techniques of government and legibility—primarily technocratic-managerialist and bureaucratic—and through the coming together of practices, events, and objects that translate constructions into material realities.[3]

Beyond imaginaries, I also turn to the practices of living with and resisting the production of risk in the lake. Julie and her flooded village illustrate the implications of urban risk for everyday lives and livelihood already transformed by other urban metabolic relations. The physical transfer of flows intersects with patterns of vulnerability shaped by past modern interventions in the lake such as aquaculture. Vulnerability to floods is understood here as a bundle of multiple conditions produced on different temporal and spatial scales, whose sources may extend beyond the city or are layered in landscape as sociomaterial assemblages.

In the following discussion I draw attention to infrastructure and its multiple paradoxes. Infrastructure is introduced to mitigate risks yet creates new risks; it aims to integrate and connect yet simultaneously fragments (Howe et al., 2016). Infrastructure is a vital lens to frame frontier making, exchange of flows, and the "politics" and "poetics" that surround such sociotechnical artifacts of modernity (Larkin, 2013). The emergence of a modern city has been historically predicated on a networked city ideal, where infrastructure integrates urban space as a whole, connecting cities to their hinterlands in the process of bringing resources closer to the urban core. But such a modern infrastructural ideal, which might have been the goal in many mid-twentieth-century urban contexts, has been shown to break down, splinter, and unbundle with urban neoliberal restructuring and subsequent urban fragmentation (Graham & Marvin, 2001). In the cities of the Global South in particular, the urban networked infrastructure ideal may have never existed in its unitary, complete form, given the colonial and postcolonial fragmentation of the urban fabric that sought to exclude indigenous and marginalized populations from urban services, who in turn produced alternative network configurations beyond the main network.[4]

Manila's network of flood control infrastructure was constructed with the modern aim of efficiently draining the city of stormwater and wastewater. Despite shifting governance contexts from colonial and postindependence to centralized authoritarian and postpolitical neoliberal, flood control remained a crucial component of Manila's urban modernity and the promise of accumulation. Similar to the modern networked infrastructural ideal, state knowledge recognized the need to expand the network of flood control beyond the city to account for the scale and scope of urban hydrology, making the urban drainage frontier legible by tracing both the upstream origins of floodwaters and possible spaces where they could be redirected. The landmark monsoon floods of 2012 and the years before and after demonstrated the breakdown of this infrastructural ideal and the limits of its design. The politics of infrastructure resurfaced and laid bare the urban metabolic connections that shape city and frontier lives.

EXTENDING THE NETWORKED CITY THROUGH FLOOD CONTROL INFRASTRUCTURE

While widespread floods occurred in Metro Manila throughout the twentieth century, the scale of the 2009, 2012, and 2013 floods was unprecedented. The material, infrastructural, and discursive origins of these contemporary floods and state responses are rooted in the early sanitary city visions of the American colonial period, the unmet modern ambitions of the postindependence state, and the state infrastructure frenzy during the Marcos authoritarian regime that enabled a radical reworking of city and lake socioecologies. Infrastructure, stormwater flows, and the spaces of the flood frontier populate this urban metabolic story.

The large-scale movement and channeling of massive material flows of water are relatively recent, mid-twentieth-century interventions drawn from much earlier plans of control. Efforts to manage Manila's floods through engineering projects date back to the early years of American colonial occupation (1898–1946) and before. Embedded within an ideal of the sanitary city—with its faith that scientific knowledge and engineering interventions can address urban problems (Melosi, 2008)—these plans were part of modern city building under a nascent but newly stabilized colonial project that sought to grant progress by imposing American ideals on a colonial setting (Anderson, 2006; Morley, 2016; Shatkin, 2005).[5]

The extensive system of urban waterways or *esteros* that earned colonial Manila its moniker Venice of the East and the problem of their drainage became the subject of engineering studies commissioned by the Manila Municipal Board in the early 1900s. In March 1905 the colonial government established the Department of Sewer and Waterworks Construction, which upon recommendations from studies prepared by American consultants crafted plans for a separate water system and a sewer system by the end of the year. Efficient conveyance of stormwater and wastes in the system of waterways through the application of scientific principles became crucial in transporting American ideals of the sanitary city to the Philippines. American city engineers assigned to the Philippines, such as Major James F. Case in 1904, actively sought engineering expertise from the United States to design Manila's own water, sewerage, and drainage systems (Manila Municipal Board, 1905b), while the expertise of a former president of the American Society of Civil Engineers was similarly solicited to review and approve their initial designs.

The use of science and engineering as interventions in the urban environment also set the American colonial project apart from three centuries of Spanish urban administration of Manila. Engineers commented on the inadequacy of the drainage network they had inherited from the Spaniards, noting they were "quite deficient for sanitary purposes, and not entirely satisfactory for the removal of storm water from the streets" (Manila Municipal Board, 1905a, p. 168). The opening observations of the superintendent of water supply and sewers submitted to the Manila Municipal Board in 1904 illustrated the novel, integrated, and scientific approach to solving the sewerage and stormwater drainage problem, which surpassed previous achievements of the Spanish colonial city administration:

> The problem of an adequate storm-water drainage system has never been fully considered. Such observations as were made by the Spanish engineers

were more or less disconnected and did not determine any point with certainty.... Before a definite design may be determined upon for all districts, observations on estero flow and groundwater levels must be taken and the results tabulated and gaugings of streams and present sewer recorded for the definite determination of the absorption and run-off coefficients. (Manila Municipal Board, 1905a, p. 103)

Over the next five years, engineers would propose and realize several localized projects for improving drainage of the *esteros* to the sea and fixing street grades and sewers. In 1910 former city engineer J. F. Case reported that "the city of Manila is now supplied with a sewer and water system, modern, sanitary, and efficient, and a capacity sufficient for its needs for many years to come" (Manila Municipal Board, 1910, p. 78). The problem of stormwater drainage became subsumed under the sewerage system design. However, it became increasingly apparent that the scale question of flooding and stormwater needed to be expanded. There was a need to "consider the general sanitary improvement of the city as a whole, making the improvement of each section a part of the general scheme and interdependent on that scheme" (Manila Municipal Board 1910, p. 85). This recognition prefigured the need for a metropolitan-level plan for drainage that exceeded erstwhile piecemeal interventions, such as building concrete river walls along the Pasig River. The modern project for a coherent, rational, and orderly city through a sanitary drainage network temporally coincided with broader urban designs for a planned modern Manila between 1905 and 1916, such as the City Beautiful–inspired Manila plan by Daniel Burnham, aspects of which were only partially realized years afterward (Morley, 2016).

These colonial engineering plans and their expansion in scale to provide drainage services to the whole of the city represent the beginnings of the modern infrastructural ideal of a centralized and standardized infrastructure network and a unitary city adopted in the Philippines. Scientific and engineering expertise that produced a knowable environment intersected with attempts to solve persistent and interlinked urban problems in a colonial city, wherein the American concern for sanitation and hygiene in the city occupied a central role (Anderson, 2007). Plans remade the space of the city through the state projects that rolled out networks from the urban core and colonial elite enclave, allowing extension and intensification of urban processes (Graham & Marvin, 2001).

It was only in postindependence Philippines, however, with the 1952 Plan for the Drainage of Manila and Suburbs, that large-scale interventions to manage and convey stormwater away from the urban core were put into a metropolis-wide plan. After almost a decade of studies and prompted by widespread floods in 1943, 1947, and 1948, the plan aimed to address two dimensions of flooding: the expansion of the built environment and the flows of water from outside the city that caused the main artery to burst its banks. The plan carried a comprehensive, holistic scheme of managing the problem on a larger, metropolitan scale, going beyond earlier piecemeal engineering projects in the urban *esteros* and stream channels. It recognized that control of the urban Pasig River flood flow required control of the upstream Marikina River in the hinterland. Taking its cue from the emergence of basin-scale management in the middle of the twentieth century (Sneddon, 2015), the plan called for a restructuring of metropolitan governance to match the hydrological scale of the flood problem. However, except for small-scale localized projects in the older sections of Manila, proposals remained unrealized until the 1970s.[6]

A confluence of several factors led to a flurry of flood control infrastructure construction and a rescaling of flood management in the 1970s. Construction of flood control structures aligned with the turn toward infrastructure of Ferdinand Marcos's authoritarian regime, showcasing the physical imprint of power and progress on the urban landscape (Lico, 2003). Marcos relied heavily on foreign borrowing through development assistance and loans to fund infrastructure projects throughout the Philippines (Boyce, 1993). This strategy addressed the primary financial limitation of the 1952 plan, estimated then to cost US$300 million and built on the emergence of the Japanese government as a source of development and technical assistance in the 1970s (Luyt, 1995; Pante, 2016).

Flood control infrastructure and softer strategies of disaster management were significant not only in building Marcos's authoritarian, modern vision of the Filipino nation as a New Society but also in shoring up popular support for the regime and quelling dissent against martial law, which Marcos imposed a few weeks after the great flood of 1972 that devastated a huge chunk of Luzon (Warren, 2013). The 1972 Manila and Suburbs Flood Control Plan, which relied on the 1952 plan, further coincided spatially with a new and emerging concept of the Metropolitan Manila region, governed by Marcos's wife, Imelda Marcos (Pante, 2016), and further centralizing urban and national state power through their conjugal rule.

The flood control infrastructure projects displayed many elements of high modernist and unitary city thinking. The 1952 and 1972 plans inherited the colonial concern for efficiently ridding the city of sanitary flows through application of scientific expertise and economic cost-benefit calculations (DPWTC, 1972; MPWTC, 1952). Both involved collection and modeling of the best available data to allow for the efficient conveyance of stormwater flows through physical interventions in the urban built environment. They also required a topological understanding of Metro Manila and its connections with surrounding areas, with the 1972 plan in particular rescaling the focus to the scale of the watershed and the hydrological regime of the Pasig River, Marikina River, and Laguna Lake basin. The two plans recognized that solutions to spare the city from inundation would require interventions in these beyond-the-city spaces by reducing or redirecting upstream flows that had caused the most disastrous flooding, notably those of the Marikina River.

As part of nation building and urban reconstruction after the extensive damage of World War II, the political context of the two flood control plans reflected tensions between local city governments and the national government over responsibility for managing floods (Pante, 2016). The centralized, authoritarian Marcos regime of the 1970s would eventually resolve these tensions by blurring the distinction between the two levels of government, best demonstrated by the creation of Metropolitan Manila as a unique urban political region in 1975. Such centralization actualized key aspects of flood control plans, through new spatial governmental bodies such as the Metro Manila Flood Control and Drainage Council, and made possible drastic interventions in the control of stormwaters. Flooding, for instance, was framed by the first Metropolitan Manila governor, Imelda Marcos, as one of the major ills that plagued the growing city and that prevented efforts toward beautification and attracting investments (Pante, 2016; Warren, 2013). Most notably, this modern vision of Manila as the City of Man bypassed the spaces of informal settlements and the urban poor, which remained poorly served by infrastructure networks. The centralization of political power enabled massive flood control infrastructure construction between 1972 and 1987, facilitated by easier access to foreign loans and suppressed opposition. In turn, it helped strengthen the legitimacy of authoritarian rule and the legacy of Marcos as builder of grand projects and the new nation. These mega-projects not only were underpinned by high modernist ideologies of efficient control of urban flows and of instilling order amid urban fragmentation, but also required the modern spatial separation of the city and the noncity.

The construction of the Mangahan Floodway, a centerpiece of Marcos's modern infrastructural vision, is illustrative of how flood control differentiates between the city as space that needs to be protected and the beyond-the-city as a space that could serve as a sink for stormwater. Carved out of marshy suburban land east of Manila in the early 1980s, the 7 km-long floodway was designed to divert stormwater flows from the Marikina River away from central Metro Manila (see map 2 in chapter 1). It was deemed the most effective solution to downstream Pasig River flooding (DPWTC, 1972; SOGREAH, 1991). The floodway is a crucial node in a network that also includes the NHCS, the Rosario Weir and its floodgates, and several pumping stations.[7] These structures—financed by foreign loans—operate synchronously to avert widespread flash flooding in Metro Manila by spatiotemporal control of stormwater flows (Liongson, 2008).

Laguna Lake plays an important role in the overall flood control design. Seventy percent of the discharge flow from the Marikina River is meant to be temporarily stored in the lake through the floodgates and the floodway. Doing so would reduce stormwater flow in the Pasig River to channel capacity. At the height of the flood events in 2009 and 2012, state engineers diverted around 4,000 m^3/s peak discharge to the lake, well beyond the capacity of the Marikina and Pasig River channels (Liongson, 2008). This diversion explains the more-than-usual flow that resulted in the swelling of the lake not seen since the great flood of 1972.

Large infrastructures as structural solutions to flooding may be effective in reducing floods in one area but often transfer risk upstream or downstream through hazard conveyance. The process of risk transference and hazard conveyance to divert flows away from an urban center is not unique to Metro Manila. Examples in other cities, such as Hanoi, Bangkok, Mexico City, and New Orleans, show contestations over where to divert flows and who will be affected (Jha et al., 2012; Lebel et al., 2009; Marks, 2019). Decisions are fundamentally political in that they rely on identifying particular places to serve as a sink and are further complicated by impacts of changing climate and rising sea levels. Politicizing risk then unearths how discourses shape flood risk, made concrete through infrastructures that affect lives and livelihoods across space.

Infrastructure becomes inherently paradoxical in that it is built to manage risk but produces new kinds of risk in the city or displaces the same risk to other spaces (Howe et al., 2016). With the construction of a flood control network, flood risk was transferred to the future by keeping the city dry at the moment and allowing continued development in areas that would have been routinely flooded otherwise. The disaster that caught Manila by surprise first in 2009 signaled a breaking down of infrastructure that accumulated risk temporally. Laguna Lake—the spatial, natural, and infrastructural solution to excessive flows of water to the city—failed to function as intended, and its dwellers were exposed to parallel risk.

The topological space of Metro Manila's flood control reflects the sanitary and networked city's preoccupation with clear and distinct separation of flooded/dry and city/noncity. Flows were to pass through the city waterways as quickly and efficiently as possible, with those that could not be channeled through urban waterways displaced to noncity sinks. Control of metabolic flows as a form of sanitary services, even as there was an eventual infrastructural breakdown later on, would become an important element in the continued functioning of the city. In this vision, Laguna Lake is the outside that makes a dry city possible. Its legibility as a particular kind of resource was produced parallel to and with the same logic as urban flood control infrastructure, emerging side by side with the modern networked city ideal. The lake becomes an infrastructure vital to the functioning of the network of flood control, in some ways resembling how nature itself becomes an infrastructure (Carse, 2012).

"A TOILET THAT COULD NOT FLUSH"

Infrastructure seeks to integrate space and time, but in the process it fragments. The flood control infrastructure network in Manila requires Laguna Lake for its hardware to function, but this transfers risk and creates a distinct othered space. Imaginaries about the lake enabled risk to take place. These narratives are part of historically evolving framings bound with urban metabolic relations between the lake and the city, while physical flows through risk transference create new discourses of a risky lake that justify further state interventions.

Territorializing Risk

In master plans for flood control infrastructure, Laguna Lake is an intrinsic part of the hydrological complex as a space for storage of excess urban stormwater. Plans identified the lake as a node in the system that comprises the Marikina, Pasig, and Napindan Rivers. Assessments saw the proposed infrastructural interventions as key in facilitating lake development schemes of harnessing resources, along with alleviating Metro Manila floods. The presence of Laguna Lake as a large body of water that can serve as retention space for water flows had been considered as far back as discussions of designing a modern water supply and drainage system under American colonial rule (Manila Municipal Board, 1905a). However, stormwater conveyance from Marikina River to Laguna Lake to mitigate downstream flooding had origins in the 1952 plan, which recognized the impacts of upstream stormwater flows on central Metro Manila floods.

The 1952 and 1972 integrated plans, as well as feasibility studies of Laguna Lake development schemes, justified the lake as a storage space because its large surface area spreads water more slowly, which creates a time lag in peak water level stage relative to the constrained capacities of urban stream channels. The plans expected the rise in lake water levels to be minimal or negligible and that the water would spread "harmlessly" over a vast area where lakeshore flooding was not considered a serious problem. The plan for the diversion channel structure, for example, suggested that "while the diversions are critical to the prevention of flooding in Manila, they are nominal when compared with the size of Laguna de Bay" (DPWTC, 1972, p. I-4). State narratives saw the lake as a space or territory that could contain excess water and where flood risks could occur (Rebotier, 2012). Such ways of seeing produce technocratic-managerialist discourses (Collins, 2009; Mustafa, 2005) that emphasize technoscientific control of biophysical processes and dimensions of risk, which can in turn be addressed through technical solution and bureaucratic management. Thus, discussions of possible solutions to Metro Manila and lake flooding after the 2009 and 2012 events revolved primarily around the provision of more or better flood control structures, and a sense of technological optimism pervaded talk of modern failures.

Seeing the lake as a smooth space, however, necessarily renders invisible social spaces and differences in the lake and justifies projects to solve issues or risk away from the more difficult, extra-local causes. Deforestation, land conversion, and land cover change in the upper Marikina River Valley are more complex issues, fueled by the increasingly capital-driven expansion of the built environment farther out into peri-urban areas (Ortega, 2016). As an intrinsic part of the operation of urban infrastructure, the beyond-the-city Laguna Lake becomes fetishized or hidden from view (Arboleda, 2016; Kaika & Swyngedouw, 2000) in the process of its imagination as a landscape of risk and in maintaining the continued operation of the city as an economic agglomeration. It performs the task of becoming infra to the flows of stormwater.

The lake provides, through diversion of stormwater, a natural, technical, and modern solution to the problem of Metro Manila flooding. Its performance as an infrastructural solution, however, is incomplete, complicated by broader problems that continue to hound managers of Laguna Lake. Seen as "a toilet that could not flush" (Dinglasan, 2012), lake problems include rapid siltation that has made the lake shallower, turbid, and highly eutrophic; reception of industrial, agricultural, and domestic wastes; and socioecological conflicts associated with the expansion of aquaculture. In turn, the state has proposed further technological interventions in the lake environment to improve its storage functions, such as the dredging of the lake bottom and urban waterways. These processes are inseparable from the lake's relations to the growth of Manila and show how risk production is inseparable from other processes that produce the lake as a modern urban resource frontier, with implications for those rendered invisible.

Mobilizing Risk

Risk is embedded in imaginaries of the lake as Manila's multipurpose resource frontier, from the first resource assessments in 1970 to water management and flood infrastructure projects of the 1980s. The city emerges from its relations with Laguna Lake as a spatial entity to be supplied with material flows and services from the lake. Risk serves as the justification for producing Laguna Lake as a space for resource production, which in turn is an opportunity to initiate parallel development interventions in the lake.

The identification of the lake where risk takes place is underpinned by two components: producing the lake as a temporary storage space to protect downstream Metro Manila through spatiotemporal control of flows and enabling hard infrastructure to allow development of the lake to supply urban demands. Both components distinguish the lake as a separate space from the city that is necessary for the city's continued functioning but not without their metabolic contradictions, such as increased risk and magnified vulnerabilities. The city/noncity differentiation also overrides the vision of the lake as a lived environment in that floods have implications for livelihoods dependent on fish production in the lake. Moreover, risk emerges as an effect of state attempts to produce an urban resource frontier, mobilized by both the state and lake dwellers for different goals.

Risk transference and lake flooding enabled further state interventions through more infrastructural investments and relocation of lake dwellers identified as residing in what were termed "danger zones" or "danger areas" (Alvarez & Cardenas, 2019). As part of the state's discursive creation of new spaces and populations at risk after the 2009 and 2012 floods, a public works official claimed that the more than seventy thousand families around Laguna Lake and more than one hundred thousand families in Metro Manila waterways living in "high risk" areas needed to be evacuated to both ease flooding and save lives (Calica, 2012). In the guise of promoting safety and managing risk, the national government allocated ₱50 billion for the systematic relocation of informal settlements along Metro Manila waterways and lakeshore sites. This has led to systematic, massive-scale "benevolent evictions" of city and lake dwellers to sprawling resettlement sites in the urban fringe, creating a new, significant driver in Manila's urban spatial expansion in peri-urban areas (Alvarez, 2019; Ortega, 2016). Inaccessible to their workplaces, poorly served by basic social services, and ironically prone to floods, these sites force many relocatees to return to their previous homes, claiming that their resettlement sites present more danger due to increased vulnerability to hazards and the lack of a source of income.

In several communities along the lakeshore, fisherfolk recount how their villages have been identified and mapped as risk- and flood-prone areas. In 2009, a few months after the Ondoy floods, President Macapagal-Arroyo signed Executive Order No. 854 revoking previous presidential proclamations that had declared sites on the northern shores of the lake and along the Mangahan floodway as socialized housing for urban poor families. Resettled from the Pasig River and other parts of the city since the 1990s, more than one hundred thousand families have settled in this now densely populated, low-lying stretch of land prone to flooding, around the area where Rizal's fictional steamship encountered unruly sandbars more than a century before (see chapter 1). Noting the need to address flooding by removing obstructions along waterways, the order proposed a plan for the relocation of residents elsewhere, which the subsequent administration carried out more systematically a few years later (Alvarez, 2019).

However, the clearing of structures and removal of at-risk populations have been uneven and selective. Driven by discourses that render certain groups of people expellable from the city and others not, many subdivisions, condominiums, and other commercial developments that encroach on waterways and the lakeshore remain untouched by demolition, while informal settlers continue to face the threat of mass evictions and are blamed for intensified risk of flood hazards (Alvarez & Cardenas, 2019). The blame of informal settlements as the cause of flooding and residents' subsequent eviction are not new despite their novel framing in the language of safety and resilience. For example, in the 1970s Marcos saw slums along waterways as an obstacle to instituting flood control programs in Manila, calling for them to be legally declared a "nuisance per se" at the height of the great flood of 1972 (Office of the President of the Philippines, 1972). This and subsequent orders facilitated and justified a series of evictions throughout the martial law period that relocated informal settlers to the peri-urban fringes (Pante, 2016).

Risk in this context is not merely materially produced or transferred to another place that is a property of the landscape but is also constructed and mobilized for particular ends. In clearing waterways and lakeshore sites to make way for new spaces of accumulation or for building urban resilience amid climate change, the state assembles practices of mapping and identifying danger zones, thereby producing discourses of risk and risky landscapes that display an exercise of state power. As I briefly narrate in the next section, lake dwellers counter state projects by also deploying notions of risk to oppose infrastructure that they believe undermines their livelihood. Risk is thus always processually reconfigured and remade within such landscapes.

RESISTANCE AND LIVELIHOOD ON A FRONTIER

Risk production and transference have material consequences for vulnerability and livelihoods of people in the extended spaces of urbanization. Frontier lives, as we have seen in chapters 2 and 3, are active in shaping their socioecological futures to make a living in a landscape of risk. Using two examples of how people challenge and produce lived realities, I turn to lake dwellers' experience of flood risk and return to this risk's intersections with urban food provisioning. The vulnerability of fishing livelihoods at the lake has been magnified by strengthening urban metabolic connections with Manila that exposed lake dwellers to greater flood hazards. They have correspondingly taken action against infrastructure as a visible conveyor of risk.

The shift to aquaculture with an explicit urban orientation in the lake heightened vulnerability of small-scale producers, particularly those already marginally positioned vis-à-vis large-scale capitalist pen aquaculture. In several fishing villages, fish producers shifted from traditional capture fisheries—livelihood activities generally unaffected by lake flooding—to cage aquaculture (see chapter 2). Unlike larger-scale and capitalist fishpen aquaculture, which tends to be more physically and financially resilient to floods and typhoons, small-scale cage aquaculture experiences the most severe impacts of extended flooding. Pen owners are also able to bounce back and reinvest in production after typhoons more promptly than cage owners (see chapter 3).

Recent floods are symptomatic of changing metabolic relations between the lake and the city, subjecting lake dwellers to greater flood risk. While bringing in more cash to the household, aquaculture simultaneously exposed fisherfolk who adopted the technology to the risks associated with producing in an environment regularly struck by typhoons and floods. Because of the fixity of enclosures in lake space, cage aquaculture allows greater control in production when compared with capture fisheries. However, it is a riskier venture because of the constant threat of damage due to wind and floods, especially when producers fail to anticipate a typhoon or when floods spill stocked fish and damage structures.

For cage producers, the arrival of the typhoon season means either reducing the number of cages in production or hauling in all nets. Those who are better financed and possess greater ability to recover and reconstruct tend to hold their ground, similar to larger-scale fishpen producers. Nets are more valuable to cage producers than fish, given that the latter can be reproduced at very low cost while the former constitute the bulk of the fixed costs (see table 2 in chapter 2). Fish escape when waters rise above overstretched nets or when floating debris, such as poles and clumps of water hyacinths, damages nets.

Small-scale cage producers lose a lot more than fishpen producers when a forecast is faulty or an unexpectedly strong typhoon hits the lake, as was the case with Typhoon Basyang in July 2010. Well-off producers, usually those who have more social and natural assets, are able to prepare and recover better after such events, supported by their ability to diversify their livelihood portfolios and spread the risk of production. With boats that have stronger engines and a bigger labor pool at their disposal, they can haul in nets before a typhoon strikes and reassemble them after it has passed. They also gain some time advantage in the mad dash for recovering washed-out nets when waters have calmed, as opposed to those who have smaller boats or those with cages that are farther from shore.

Recovering from typhoons involves reinvesting in nets and poles, which requires a significant sum of money, usually the equivalent of at least three-quarters of cage construction costs. Cage producers use the phrase "going back to zero" to refer to this situation, further burdened by the need to pay annual leases to the LLDA. Access to credit from kin and through microfinance is important in this recovery. Producers are forced to downsize production, and some work temporarily as laborers for owners of other cages. Newer entrants to aquaculture and poorer households are especially impacted. As a small-scale producer noted, his limited capacity to make his cages more hazard-proof led to loss of stocked fish during the 2009 typhoon, which was weak in terms of wind speed but brought severe damage due to raised water levels.

Cage aquaculture producers see recurring flood and typhoon damage as preventing them from expanding production. For them, the occurrence of typhoons is something uncontrollable and inevitable, part of and intrinsic to cage aquaculture in the lake. Nonetheless, these hazards are more predictable, and preparations are possible. The hauling in of nets at the beginning of the typhoon season and other similar practices involve being strategic in timing (*madiskarte*). Others harvest or sell their fingerlings or fish early when they hear of an approaching typhoon. Saving enough money from each sale to prepare for the damage of future typhoons is also important in recovery. Vulnerability, as illustrated by cage producers in Laguna Lake, is more than just a static condition of certain groups of people; instead it is dynamically produced through the intersections of structures such as aquaculture and flood control development projects and multiple social and power relations and practices among lake producers.

Lake dwellers recognize their links with the city through infrastructure and its consequences for their livelihoods, such as how these structures control flows vital to fishing at their expense and for the benefit of water consumers in Manila. Identified as a silting space for stormwater storage, the lake has been subject to several state attempts to increase its ability to hold more water and improve its flood control functions. Many of these efforts—including dredging and road dikes—were heavily debated and opposed by lake producers because they threatened fisheries' productivity (Cinco, 2011).

Infrastructure becomes a rallying point in calls for better lake socioecologies. Because of the material and symbolic importance of flood control structures, these are seen as the root cause of lake problems such as flooding and poor fish productivity. In both 2009 and 2012, fisherfolk and other lake-based groups called for the opening of the gates of the NHCS to allow diverted water to flow out to Manila Bay and reduce lake flooding (Mayuga, 2012). Since the construction of the Mangahan Floodway and NHCS during the authoritarian Marcos regime, there has been a general sentiment among lake dwellers that these are responsible for recurrent floods.

The precedent for resistance against infrastructure was set in the 1980s, when the NHCS was constructed in a Pasig-Napindan junction to solve the problem of saline intrusion and high nutrient load. Drawing from experiential knowledge of lake ecology, lake producers opposed the NHCS on the grounds that it limited the vital saline flux that seasonally revived the fisheries (see chapter 1). Historically in conflict over aquaculture's enclosure of the lake commons, groups of fisherfolk and the politically well-connected pen aquaculture producers joined forces in 1985 to pressure the state to undo the NHCS's functioning and allow unregulated flow of water, returning both saltwater and other nutrients to the lake.

In a series of actions that reflect the fluid connections between the lake and the city, fisherfolk mobilized a parade of boats to the presidential seat in Malacañang along the Pasig River and persuaded the state to prevent the operation of the structure as saline control. For some lake dwellers, this history of mobilization against the NHCS parallels more recent opposition to proposed road dike infrastructure seen to cause widespread impacts on livelihood and risk. As one fisher in his sixties recalled:

> We held a rally, a motorcade. We went to Malacañang [the presidential seat] riding boats but we were not allowed inside to talk to the president. This is what I want to happen again to show the government that we are against this road dike. What happened back then was that we contributed a peso each so that we could try to blast the [NHCS]. . . . If some of the people I know back then were still alive today, [they would oppose this proposed road dike]. The grandfather of one of my friends once told him that there will come time someday when the lake will turn red. Maybe this is the time. The construction of the road dike is a sign of that. Lives might be shed.[8]

Fisherfolk living near the proposed site of the road dike have also deployed notions of risk to oppose it, arguing that it will magnify, rather than mitigate, flooding in their specific area. In the case of both past and present controversies, the construction of large, visible infrastructure became the target of opposition for its ability to shape lake socioecologies and livelihoods. Opposition to infrastructure showed not only contestations between local knowledge and state technocratic-scientific discourse of lake ecology but also resistance to the production of urban frontiers.

CONCLUSION

In this chapter, I framed the monsoon floods of 2012 as symptomatic of the urban metabolic processes that enroll frontier spaces through the production of landscapes of risk mediated by infrastructure. Urban risk embodies the tensions and relations between the city and frontier, which enable the possibility of a dry city, a crucial precondition for urban growth and accumulation. The contemporary moment captured in this chapter shows how physical control of flows is underpinned by layered territorial imaginaries from the past and the work of infrastructural maintenance situated in the city. The effects of this control extend beyond fragmentation in the city, intersecting with lives and livelihoods in the urban frontier on the edge.

Manila's flood control network, like other infrastructure, is in many ways paradoxical (Howe et al., 2016). Its construction embodies visions of progress, control, and unity, yet it also symbolizes modernity's fragmentation and failure to deliver on its promise. The symbolic power of its imprint on the landscape legitimizes the state and attempts to build a modern nation and city, but it often fades into the built environment, becoming visible and contested again when it fails. The flood control network's visibility is necessary for its political and poetic effects. Infrastructure aims to keep the city safe and dry, but this requires channeling harm elsewhere. It does not merely bring resource flows to the city but also produces and reproduces uneven landscapes of risk that become political targets in addressing urban metabolic inequality.

Resistance to infrastructure in Laguna Lake, such as the NHCS, signifies a rejection of the privileging of technocratic-managerialist knowing of the lake. Yet the future trajectories of lake resistance and politics surrounding infrastructure remain uncertain, given that the lake itself is performing the work of and becoming infrastructure. Turning to how beyond-the-city spaces are enrolled in urbanization extends questions of urban politics and resistance. Beyond contestations of imaginaries of the urban frontier, urban politics shapes the material geographies that configure socioecological futures and conflicts around them. Distinguishing urban politics in the noncity reveals the fetishism of landscapes that makes the operation of cities possible while emphasizing the processual character of urban politics within and beyond the city.

EPILOGUE

Mutable Frontiers, Metabolic Futures

[The government officials] kept on telling us that the lake is dead. But what do they know about the lake? They only have ballpens in their hands; we hold fishing nets.

ERNESTO, *fisher, August 31, 2015*

IN THIS BOOK, I turned to the ecologies on the urban edge to narrate how urbanization is a frontier-making process that enrolls an assemblage of natures, landscapes, infrastructures, and peoples within and beyond the city in search of solutions to persistent urban resource problems. I showed how paradoxical socioecological transformations in urban provisioning between Manila and Laguna Lake required the practical and imaginative work of making and maintaining. These metabolic processes have been uneven, contested, political, and driven by multiple logics of state, capital, and everyday livelihood making of people inhabiting urbanizing environments.

Frontiers are mutable, mobile, and temporal entities that emerge, vanish, move on, and return (Cons & Eilenberg, 2019; Tsing, 2005). By way of conclusion, I revisit these questions of temporality in urban frontiers. What happens to a frontier indelibly transformed to deliver urban metabolic resource flows? What socioecological futures (Braun, 2015) are imagined for a space on the urban edge if and when the commodity frontier moves on? How do imaginaries of ruination reconfigure resource frontier landscapes, and to what political effect? Shifts in lake governance since 2010 provide glimpses of future lake trajectories.

During a visit to a lakeshore community in the aftermath of flooding in 2013, President Benigno Aquino III promised an infrastructural solution to floods that have increasingly worsened along Laguna Lake and Metro Manila (Cinco, 2013). Promoted as his administration's biggest public-private partnership project, the Laguna Lakeshore Expressway Dike (LLED) sought to solve flooding along the western shores of the lake while easing traffic congestion on Manila's southeastern edge (Burgonio, 2014). The construction of the 43 km road dike also entailed reclamation of 700 ha of land from the lake, paralleling a similar reclamation rush in Manila Bay for urban and real estate development. Key Filipino corporations that have profited from a recent urban real estate investment turn submitted bids for the 123-billion-peso project in 2015. Opposition to the proposed dike emerged simultaneously with the project's development. Groups of small-scale fisherfolk and aquaculture producers expressed concern about negative impacts on livelihood and constrained lake access. Residents worried that flood risk would intensify on their shores in the process of keeping other sections of the lake flood free. Older fisherfolk who had lived through the height of resource conflicts decades ago wondered if this was the 1980s all over again.

Put into the historical context of urban resource frontier making, the emergence of the LLED as a mega-infrastructural intervention is nothing new despite the project's novel neoliberal flavor. The state has always prioritized certain metabolic flows over others, even when making claims about win-win solutions and keeping the multi-use resource character of the lake through resource governance. While the LLED project was eventually shelved when bidders backed out in 2016 due to financial infeasibility (Camus, 2016), it reflected the changing dynamics of the lake's place as a resource frontier.

Trajectories of Laguna Lake's socioecological futures may be interpreted in a number of ways. At one end, neoliberal, corporate real estate–driven capital accumulation, configured by global capital flows and manifested in intensified landscape change at the urban edge, signals the displacement of older configurations of agrarian/urban capitalism entrenched at the lake since the 1970s. Like Manila Bay to its west, reclamation of Laguna Lake to produce more land has become more attractive to capital as a speculative form of urbanism than the crisis-ridden, water-based fish production that it is easing out. At the other end, aquaculture has simply exhausted its surplus-making capabilities and will give way to other modes of accumulation. The lake is increasingly reconfigured to deliver new forms of appropriation in a space where older agrarian capitalist formations are less able to produce sustained accumulation. New state and scientific arrangements and discourses have emerged to enable transformations that channel particular metabolic flows and transform environments. The success of capitalist aquaculture was in its ability to harness the high ecological surplus provided by the lake and its people, but exhaustion of these arrangements has failed to produce further rounds of accumulation. Processes that seek to reconfigure lake natures create novel arrangements of human/nonhumans that can deliver more commodification, manifest in conflicting metabolic flows and future visions of the lake, threatening the very pioneering activity that opened up the lake to capital's intrusion.

Yet imaginaries of ruination—decline, death, and decay—have always been a key feature of resource frontiers and infrastructures as their productive capacities are exhausted or rendered obsolete (Cons & Eilenberg, 2019; Howe et al., 2016). The imagination of the lake as dying, dead, polluted, and degraded continues to circulate as a truism: the silted, murky brown lake heavily transformed for decades by human activities has become nothing more than a utilitarian landscape of resource production. As with past state interventions in the lake, the framing of Laguna Lake as dead or dying and that fewer people depend on the lake for daily sustenance serves to justify megaprojects and make these interventions less contentious and more acceptable to the public. In contrast to and despite attempts to silence their narratives or erase their livelihood, fisherfolk and lake residents continue to emphasize the lake as lively and life giving in mobilizing their opposition. Ernesto's quote at the opening of this epilogue, a remark he made after attending a government consultation meeting on the proposed LLED project, captured these differences in state and fisherfolk knowledge and dependence on the lake. It emphasized the importance of the situatedness of working with the lake in understanding its condition and its fate.

On the other end of the spectrum, Laguna Lake as a landscape of ruins of past resource production serves a political purpose beyond legitimizing further infrastructural interventions. The year 2016 also marked an apparent shift in national politics and environmental governance, with the election of President Rodrigo Duterte through a largely strongman populist platform. Rather surprisingly, he immediately turned his attention to the lake as an environmental problem that can be solved through an exercise of political will and authoritarian interventions reminiscent of Marcos decades before him. His rhetoric departed drastically from the liberal, good governance models of Aquino and others who have built mechanisms of postpolitical consensus into resource management. Blaming the elite for the marginalization of the poor, Duterte played up past antagonisms, albeit framed in highly simplistic dualisms, between rich fishpen owners and displaced fisherfolk.

With a strongly populist undertone, Duterte promised to bring the lake back to a past pristine state of bounty. This he sought to do through a reordering of the lake through the old language of rationalization. Bringing order back to the lake and taking power from elite fishpen owners would restore the lake to its past condition, allowing apparently less exploitative activity such as ecotourism to take root (Gamil, 2016). Understandably, fishers and aquaculture producers were concerned about the implications for their livelihoods of such plans, with fisherfolk groups supporting the needed check against pen expansion but opposing early plans to totally remove fishpens from the lake. Many fishers have built multiple connections and relations with the fishpen economy, and their complete removal from the lake would significantly undermine their livelihoods. Fishpen producers, meanwhile, continued to emphasize their role in urban fish provisioning and food security. A few years after Duterte unsettled lake politics and after a number of publicized demolitions and military interventions in 2017, the status quo seemed to have returned, and the pristine lake imaginary was sidelined.

The very same administration has proposed new projects under its banner "Build, Build, Build" program centered on the construction of mega infrastructure to solve many of Manila's metabolic problems (Mouton & Shatkin, 2020). Providing water for Manila has again become a contentious issue, with threats of a parched city during exceedingly dry years. Turning to frontier spaces on the edge and beyond for solutions, proposals include extracting more Laguna Lake water for urban domestic use and the Chinese-funded construction of dams in the Sierra Madre Mountains to the east of the city. These new projects have been contentious because they would displace indigenous peoples from their ancestral lands in the mountains and pose threats to the fisheries' livelihood in the lake. Even with these new sources of water supply, access to water, food, and services remains splintered according to class and fragmented across space in Manila and surrounding regions without any fundamental shifts in governance strategies for addressing uneven urban development. Urbanization continues to shape metabolic inequalities in the city around various emerging urban forms and old and new urban frontiers despite apparent changes in state governance rhetoric.

One promise of an urban metabolic lens is that it extends the stuff and sites of politics. Tracking metabolic flows takes us to various sites of dispossession and inequality and how some forms of metabolism are valued over others. It entails including more political actors, potential sites of intervention, and ways of seeing (McFarlane, 2013; McFarlane & Silver, 2017). The urban metabolism of fish brings into focus the plight of displaced fisherfolk at the lake alongside the plight of urban workers at the fish port. Socioecological change in the lake has magnified risk for urban poor consumers in a way that is similar to flood infrastructure risk magnified for aquaculture cage producers. Fish markets, household kitchens, aquaculture farms, the flood control operations center, points of access to infrastructure, and so on are sites where power is produced and contested and where potential alternatives are built through everyday encounters and incremental change (Lawhon et al., 2014), including both the radically transformative and the seemingly banal.

Yet despite urbanization rolling out farther into the spaces beyond the city, it should be emphasized that political imaginaries and socioecological futures of people in these various sites are neither completely oriented toward the city nor subsumed solely under urbanization imperatives (Paprocki, 2020). Frontiers are urbanizing, but these spaces cannot be reduced to only this process. Edges, after all, are sites where multiple eventualities emerge and diverge as urbanization rolls out temporally with incomplete and undecidable trajectories. The politics of edginess here extends beyond merely recognizing and seeing urban metabolic connections. It also invites viewing urban edges as vital grounds for forging political possibilities as people attend to their livelihood needs through diverse rhythms of practice (Simone, 2019), such as in the patchwork, leftover contiguities of the often overlooked urban frontiers.

Narratives of metabolic conflicts and the political ecologies of how certain metabolisms are privileged over others do not aim to end with a reification of material flows. Nor do they merely seek to locate urban politics in either the city, the frontier, or their edges. Rather, these narratives foreground the processes and configurations that produce and reinforce uneven and unequal relations, spaces, and ecologies. Talk of metabolic politics is not a mere abstraction but is always concrete and situated in everyday life as people participate in the reproduction of the existing or the production of a different socionatural order. It signals the possibility of transforming existing unjust relations by mixing labor with nature, as in Ernesto's understanding of the lake as living and life giving mediated through his work of fishing. Metabolism is the coproduction of nature and society, one that is always situated in practical activities of the everyday.

NOTES

INTRODUCTION: FRONTIERS OF URBANIZATION

1. Metro Manila is also the National Capital Region (NCR) and consists of seventeen local government units, with a 2015 population of thirteen million. It is included in the Greater Manila Area (population twenty-five million), along with the adjacent provinces Cavite, Laguna, Rizal and Bulacan, and extends farther out into the nearby regions of Calabarzon and Central Luzon as the Mega Manila Urban Region (population forty million).

2. The City of Manila is the official capital of the Philippines and is often considered the densest city in the world in terms of population per square km. It contains the old urban core, founded as a trading post in the thirteenth century, and became a Spanish colonial city in the sixteenth century (Morley, 2018).

3. Laguna de Bay, translated from Spanish as Lake of Bay (also Bae or Ba-i) after a town on its southern shore, is the colonial and official name of the lake. Laguna Lake, an inaccurate translation, was used by the government with the emergence of its regulatory body, the Laguna Lake Development Authority. I use Laguna Lake throughout the text instead of Laguna de Bay to emphasize this aspect of state resource production and legibility expressed in a rather redundant, Anglicized name.

4. The ecological, economic, and cultural differences of producing and consuming these three species will become more evident in subsequent chapters. Of the three, milkfish has the longest history of consumption as an indigenous species that has long been cultivated in fishponds elsewhere in the country. Tilapia and bighead carp were both introduced in the mid-twentieth century from Africa and China, respectively, but have since found widespread appeal especially to lower- and middle-income consumers.

5. Fisherfolk is a collective term used to refer to a wide range of people who make a living from fisheries, employing various gear and arrangements. Chapter 1 provides a more extensive discussion of these activities and identities in relation to capture fisheries in the lake.

6. For comprehensive overviews of urban metabolism from a systems-based perspective, see for example Baccini and Brunner (2012), Ferrão and Fernandez

(2013), Fischer-Kowalski (1998), and Zhang (2013). For classic and recent case studies of individual cities and multicity comparative studies using a variety of urban metabolism quantification tools, see also Barles (2007), Huang (1998), Goldstein et al. (2017), and Warren-Rhodes and Koenig (2001).

7. Both the metabolic rift theory school and urban political ecology draw from Karl Marx's conception of metabolism as mediating nature-society relations. The former tracks nature-society ruptures as a result of capitalist expansion cleaving these relations (Clark and Foster, 2009; Clausen and Clark, 2005; Foster, 1999), while the latter focuses on hybridity in urban ecological transformations (Gandy, 2004; Heynen et al., 2006; Swyngedouw, 1996, 2006). Moore (2011, 2015), Davies (2019), and Napoletano et al. (2019) present recent ontological debates about Marxist metabolism and questions of nature-society dualism.

8. See UPE work on water (Cousins, 2017; Delgado-Ramos, 2015; Gandy, 2003; Kaika, 2004; Loftus, 2006; March, 2015; Swyngedouw, 2004) and other flows (Demaria & Schindler, 2016; Guibrunet et al., 2017; Lawhon, 2013; Marvin & Medd, 2006; Yates & Gutberlet, 2011).

9. Outside of an urban political ecology approach, the relationship between urbanization and food systems has been examined in the context of sustainability and pressures on resource transformations in foodscapes, foodprints, or foodsheds linked through food flows and supply chains to increasing populations and incomes and diversifying tastes and lifestyles in cities (Barles, 2007; Goldstein et al., 2017; Seto & Ramankutty, 2016; Wegerif & Wiskerke, 2017).

10. The rich body of work on the geographical lives of commodities encompasses the commodity chain and related approaches (value chain, *filiere*, global production network), as well as the social organization of commodity production and exchange in the fields of economic geography and economic sociology (Bair, 2009; Bernstein & Campling, 2006; Gereffi et al., 1994; Watts, 2005). It also includes cultural economic and material cultural approaches to the material-discursive constitution and consumption of commodities (Appadurai, 1986; Cook, 2004).

11. For the key arguments of the Lefebvrian-inspired planetary urbanization concept and its challenges to the epistemological category of the urban, see Brenner and Schmid (2015). Empirical works that have taken up some of the approaches of planetary urbanization include Arboleda (2016), Kanai (2014), Kanai and Schindler (2019), and Rice and Tyner (2017). Critical responses and debates about the usefulness of the concept have similarly emerged from feminist, queer, and postcolonial perspectives, among others (Buckley and Strauss, 2016; Jazeel, 2018; Oswin, 2018; Reddy, 2018; Roy, 2016a).

CHAPTER 1. BIRTH OF A CONVENIENT FRONTIER

1. Jose Rizal (1861–1896) was a Filipino writer, doctor, and nationalist. His two novels and his execution inspired the Philippine revolution against Spanish colonial rule.

2. Ishmael Bernal (1939–1996) was a national artist of the Philippines for Film. A number of his films made during Ferdinand Marcos's authoritarian rule depicted the lives of ordinary urban and rural Filipinos that contrasted with the utopian visions of Marcos's "New Society".

3. After being elected to the presidency in 1965 and 1969, Marcos declared martial law in 1972. The rise of Marcos and the particular political economy of his regime—populated by technocrats, cronies, and the military—have been extensively documented elsewhere (see, among others, Anderson, 1988; McCoy, 2017; Noble, 1986; Tadem, 2013).

4. *Balut* is often peddled and consumed as a nighttime street food in both urban and rural Philippines. The making of *balut* is said to have been introduced by Chinese traders who settled along Laguna Lake, developing an industry in lakeshore towns like Pateros (Alejandria et al., 2019).

5. Milkfish produced in the fishponds of Central Luzon was brought to the city through rail transport in the early twentieth century (Herre & Mendoza, 1929).

6. The lake's ecology does not support the natural reproduction of milkfish.

7. Tilapia genetic improvement has a long history and has revolutionized tilapia production globally. Aside from the GIFT project introduced in the 1980s (Khaw et al., 2008), the Genetically Male Tilapia project, which improves fish growth by producing all-male progeny through chromosome manipulation (Acosta et al., 2006), also developed in the Philippines in the 1990s.

8. Projected water demand for Metro Manila is expected to increase to 6,950 MLD in 2025 from the current water demand of 5,600 MLD (Tabios, 2019). The Angat Reservoir currently supplies 75–80 percent of water demand, with Laguna Lake supplying 4 percent and groundwater sources, 14 percent. Tabios (2019) estimates that Laguna Lake can provide as much as a third of this total urban demand.

CHAPTER 2. ENCLOSING A COMMODITY FRONTIER

1. See Campling (2012) and Ertor and Ortega-Cerda (2019) for similar examples in fisheries.

2. This LLDA–local government conflict was finally resolved in 1995 when the Supreme Court ruled that the LLDA held the mandate to grant permits and collect fees from lake aquaculture.

3. The cities (formerly towns) of Navotas and Malabon were incorporated into Metropolitan Manila or the National Capital Region in 1975. They are located along the northeastern coast of Manila Bay and have developed as the fishpond and fisheries center of the region.

4. Bulacan fishpond owners to the north of Malabon and Navotas also joined this initial rush, owing to their access to milkfish fry and fingerlings, which they produced in their fishponds and farmed for grow-out in the lake. Milkfish fingerlings are still transported to the lake through boats (*pituya*) from the shores of Bulacan through Manila Bay and the Pasig River to the lake.

5. It was common for various fishing gear, especially with the advent of diesel-powered motor boats in the 1950s, to follow the fish throughout the lake regardless of territorial jurisdiction. With the passage of the 1998 Fisheries Code, regulation of fishing activities in the lake also depended on local government enforcement, which was often uneven and varying.

6. Interview, May 5, 2012. All interview respondents' names in the book are pseudonyms.

7. Interview, May 8, 2012.

8. Rey, interview, May 8, 2012.

9. Jomar, interview, May 1, 2012.

10. For a history of rural insurgency and peasant revolt in the Philippines, see Franco and Borras (2007), Kerkvliet (2002), and Lara and Morales (1990).

11. Empirical work on these topics is extensive in the political ecology, agrarian change, and coastal/inland fisheries literature (e.g., Belton & Little, 2008, 2011; Belton et al., 2012; Bene et al., 2016; Das, 2014; Ertör & Ortega-Cerda, 2015; Hall, 2004; Kelly, 1996; Pant et al., 2014; Paprocki & Cons, 2014; Stevenson & Irz, 2009; Stonich et al., 1997; Veuthey & Gerber 2012).

12. Kalinawan is located on the western shore of the peninsula that separates the West Bay from the Central Bay. While technically on the mainland, the village is considered part of "Isla" (Talim Island) because like the island villages, it is only accessible by boat. The narrow flat shoreland of the village quickly makes way for the steep, sharp, and rocky hills that characterize most of Isla. Most of the 2015 population of 2,062 is engaged in small-scale aquaculture, with a few households engaged in other fish-related livelihoods, charcoal making, swidden agriculture, and urban off-farm work.

13. Cage aquaculture, more precisely, is production of fish through several cages, with enclosures averaging half a hectare, both for nurseries and grow-out. Nurseries produce fingerlings, while grow-out produces fish for household consumption or sale to the market.

14. Interview, June 26, 2012.

15. Thousands of tilapia fry that spawn from adult breeders are regularly graded and transferred to other cage enclosures every two weeks to avoid crowding and cannibalism.

16. Many villagers began engaging in full-time and seasonal off-farm work in the *kati* or mainland since the 1970s and 1980s. A few have also worked as overseas workers, particularly since the 1970s under the state policy of exporting labor.

17. Owing to the difficulty of transporting construction materials, building concrete houses cost more in Isla. The ability to afford such construction becomes an indicator of wealth in the villages.

18. Jess, interview, June 13, 2012.

19. Noy, interview, June 14, 2012.

20. Tonyo, interview, June 20, 2012.

21. This form of partnership would have different share agreements, for example, 70–30 in favor of the financer.

22. A husband-wife household head can own and manage one or several cages and employ hired labor or household labor (usually sons or other male relatives). Larger cages employ stay-in laborers, who are paid a fixed monthly wage apart from food, with some getting shares from harvest profits. Smaller cages rely more on household labor, often unpaid but sometimes waged. Hiring of extra seasonal labor is necessary for the high labor requirements of tasks such as repairing nets and harvesting and delivery of fish.

23. Manny, interview, May 14, 2012.

24. Jose, interview, June 19, 2012.

25. Navotas, different from the Metro Manila city of Navotas where the Navotas Fish Port Complex is located, is situated at the northernmost tip of Talim Island and is separated from the mainland by Diablo Pass, the deepest portion of the lake. Of the population of 3,262 in 2015, most are engaged in capture fisheries through various gear (around 65 percent of households) as well as small-scale aquaculture (10 percent) and fish trading (5 percent). Fishpen ancillary work and fish drying, alongside charcoal making, swidden agriculture, and off-farm work comprise the other livelihood occupations of villagers.

26. If caught by the Bantay Lawa (Lake Watch), boats are seized and confiscated, and crew members are temporarily jailed until bailed out by the boat owner.

CHAPTER 3. AN UNRULY FRONTIER

1. This chapter draws from a variety of approaches to the question of nature, including those that emerged in agrarian political economy, ecological Marxism, political ecology, new materialism, and others. Rather than rehearse or resolve debates and contentions about ontology, intentionality, agency, and similar concerns, I present narrative diversity in demonstrating the place of nature in Laguna Lake frontier making.

2. Interview, November 8, 2012.

3. Terry, interview, May 7, 2012.

4. Subsequent revisions to the ZOMAP were passed in 1999 and 2003.

5. Tonyo, cage producer, interview, June 20, 2012.

6. Francis, fishpen operator, interview, July 6, 2012.

7. Joel, interview, November 8, 2012.

8. Sonny, interview, May 15, 2020.

9. Interview, November 8, 2012.

10. Gardo, interview, May 8, 2012.

11. Roger, interview, May 8, 2012.

12. Jeric, interview, May 1, 2012.

13. Bighead carp and tilapia are produced artificially using hatcheries in tanks onshore and nets in cages in the lake, respectively. Milkfish fry and fingerlings, meanwhile, are sourced from outside the lake, mostly from the Central Luzon province of Bulacan.

14. Joel, interview, November 8, 2012.
15. Interview, May 8, 2012.

CHAPTER 4. CHAINS OF URBAN PROVISIONING

1. Work on commodity chains (Bair, 2009; Bernstein & Campling, 2006; Gereffi et al., 1994; Hudson, 2008; Watts, 2005) and value chains (Bair, 2005; Belton & Little 2011; Bene et al., 2016; Kaplinksy, 2000) shows diverse approaches to mapping movement of commodities and the networks and circuits of relations produced in the process.
2. Aquaculture in Taal Lake, Batangas, located a few kilometers southwest, parallels many of the ecological problems of Laguna Lake, most notably recurring fish kills as a result of the overfeeding of fishcages.
3. The most common fishing gear in the commercial sector is ring nets, trawls, purse seines, and bagnets. Tuna, sardines, and round scad are the primary types of fish caught by commercial fisheries, with tuna as a primary export product of the Philippines at around 100,000 MT exported annually to destinations such as Japan, the United States, and Germany.
4. Bryan, fishpen operator, interview, November 8, 2012.
5. Sonny, trader-transporter, interview, May 14, 2012.
6. The practice has also been observed in other smaller fish auctions elsewhere in the Philippines (Turgo, 2012).
7. Charlie, interview, October 15, 2012.
8. Larry, interview, November 14, 2012.
9. Ramil, interview, November 14, 2012.
10. Larry, interview, November 14, 2012.
11. Jason, interview, November 14, 2012.
12. Jason, interview, November 14, 2012.
13. Larry, interview, November 14, 2012.
14. Mon, interview, November 14, 2012.
15. Larry, interview, November 14, 2012.

CHAPTER 5. BIOGRAPHIES OF FISH FOR THE CITY

1. Cities in Southeast and South Asia such as Bangkok, Kolkata, Phnom Penh, Ho Chi Minh City, and Hanoi display parallel challenges of producing farmed fish for urban consumption amid ecological change associated with urban development; encroachment; and conversion of wetlands, rice farms, and coastal zones that support peri-urban aquaculture, including regionally specific wastewater-fed aquaculture and integrated agriculture-aquaculture systems (Bunting & Little, 2015).
2. Similar arguments of food's role in social reproduction have been made in a few other works (see Breitbach, 2007; McMichael, 2003; Shillington, 2013; Strauss, 2013).

3. Artificial reproduction of bighead carp differs from tilapia considerably and involves more complex techniques and equipment. Bighead carp fry is produced through induced spawning in inland tanks, wherein bighead carp sire sperm is manually mixed with dam eggs. Hatched fry are then graded and transferred to other tanks before being transferred to nurseries in the lake until they grow to fingerling size.

4. Interview, March 4, 2012.

5. An invasive species of tilapia, the black-chin tilapia (*Sarotherodon melanotheron*), that had found its way to Laguna Lake and Manila Bay is coincidentally known locally as *tilapiang gloria*, named after the fish's resemblance to the president (Ordoñez et al., 2015).

6. Distribution of households by socioeconomic class in Metro Manila in 2008, according to the Philippine Statistics Authority, is upper (A/B) classes, 4.7 percent; middle (C) class, 23.5 percent; lower (D) class, 45.1 percent; and extremely lower (E) class, 26.7%.

7. Mon, interview, November 14, 2012.

8. Mel, fish retailer, interview, November 1, 2012.

9. Maria, fish consumer, interview, November 28, 2012.

10. Lucille, interview, December 2, 2012.

11. Eddie, interview, December 3, 2012.

CHAPTER 6. INFRASTRUCTURES OF RISK

1. The 2012 *habagat* (southwest monsoon) flood was the result of torrential rains August 1–8, 2012, caused by circulation associated with Typhoons Gener and Haikui. More than 1,000 mm of rain fell on the city in a span of seventy-two hours. Rainfall amounts were comparable to Tropical Storm Ondoy in 2009, which brought a record of nearly 500 mm of rain in twelve hours in some sections of Metro Manila, affecting nearly 80 percent of urban households and causing damage to infrastructure estimated at nearly $100 million (Abon et al., 2011; Heistermann et al., 2013; Nilo and Espinueva 2011).

2. Anna, interview, October 23, 2012.

3. See extensive work on diverse approaches to risk production and the city (Alvarez & Cardenas, 2019; Blok, 2016; Collins, 2009, 2012; Dooling & Simon, 2012; Gustafson, 2015; Marks, 2019; Ranganathan, 2015; Simon, 2014; Walker et al., 2011; Yamane, 2009; Zeiderman, 2012).

4. These arguments have been proposed by a number of urban infrastructure studies (Coutard, 2008; Coutard & Rutherford, 2015; Furlong & Kooy, 2017; Gandy, 2006; Kooy & Bakker, 2008).

5. The Americans turned to Manila's urban problems after devoting the first few years of colonial occupation to military operations during the bloody Philippine-American War (1899–1902).

6. These projects included the Blumentritt Intercepter in 1955, considered the first major project under the plan, as well as two pump stations, five floodgates,

and more than 8 km of Pasig River flood walls, all constructed by 1967 (Pante, 2016).

7. The plans also included the Paranaque Spillway, which was supposed to connect Laguna Lake with Manila Bay through a channel along a narrow portion of southern Metro Manila. However, the project was never constructed because of the high estimated costs of the project and the increasing density of people living along the proposed path of the spillway. The floods in the 2010s revived interest in the spillway and another project that aimed to channel water from the lake through the Sierra Mountain Range to the east and into the Pacific Ocean (Dinglasan, 2012).

8. Lito, interview, August 31, 2015.

REFERENCES

Abarra, S. T., Velasquez, S. F., Guzman, K. D. D., Felipe, J. L. F., Tayamen, M. M., & Ragaza, J. A. (2017). Replacement of fishmeal with processed meal from knifefish *Chitala ornata* in diets of juvenile Nile tilapia *Oreochromis niloticus*. *Aquaculture Reports*, 5, 76–83.

Abon, C. C., David, C. P. C., & Pellejera, N. E. B. (2011). Reconstructing the tropical storm Ketsana flood event in Marikina River, Philippines. *Hydrology and Earth System Sciences*, 15(4), 1283.

Acosta, B. O., Sevilleja, R. C., & Gupta, M. V. (2006). *Public and private partnerships in aquaculture: A case study on tilapia research and development*. Penang: WorldFish Center.

Adduci, M. (2009). Neoliberal wave rocks Chilika Lake, India: Conflict over intensive aquaculture from a class perspective. *Journal of Agrarian Change*, 9(4), 484–511.

Aguilar, F. V., Jr. (1989). The Philippine peasant as capitalist: Beyond the categories of ideal-typical capitalism. *The Journal of Peasant Studies*, 17(1), 41–67.

Akram-Lodhi, A. H., & Kay, C. (2010). Surveying the agrarian question (part 2): Current debates and beyond. *The Journal of Peasant Studies*, 37(2), 255–284.

Aldaba, V. C. (1931a). The dalag fishery of Laguna de Bay. *Philippine Journal of Science*, 45(1), 41–60.

Aldaba, V. C. (1931b). Fishing methods in Laguna de Bay. *Philippine Journal of Science*, 45(1), 1–28.

Aldaba, V. C. (1931c). The kanduli fishery of Laguna de Bay. *Philippine Journal of Science*, 45(1), 29–40.

Alejandria, M. C. P., De Vergara, T. I. M., & Colmenar, K. P. M. (2019). The authentic balut: History, culture, and economy of a Philippine food icon. *Journal of Ethnic Foods*, 6(1), 1–10.

Alix, J. D. (1976). Survey of fish catch landed and unloaded at the Navotas Fish Landing and Market Authority in Navotas, Metro Manila. *Fisheries Research Journal of the Philippines*, 1(2), 50–56.

Alvarez M. K. (2019). Benevolent evictions and cooperative housing models in post-Ondoy Manila. *Radical Housing Journal, 1*(1): 49–68.

Alvarez, M. K., & Cardenas, K. (2019). Evicting slums, "building back better": Resiliency revanchism and disaster risk management in Manila. *International Journal of Urban and Regional Research, 43*(2), 227–249.

Anderson, B. (1988). Cacique democracy and the Philippines: Origins and dreams. *New Left Review, 169,* 3–33.

Anderson, W. (2006). *Colonial pathologies: American tropical medicine, race, and hygiene in the Philippines.* Durham, NC: Duke University Press.

Angelo, H., & Wachsmuth, D. (2015). Urbanizing urban political ecology: A critique of methodological cityism. *International Journal of Urban and Regional Research, 39*(1), 16–27.

Appadurai, A. (1986). *The social life of things: Commodities in cultural perspective.* Cambridge: Cambridge University Press.

Arboleda, M. (2016). In the nature of the non-city: Expanded infrastructural networks and the political ecology of planetary urbanization. *Antipode, 48*(2), 233–251.

Arcilla, C. A. C. (2018). Producing empty socialized housing: Privatizing gains, socializing costs, and dispossessing the Filipino poor. *Social Transformations: Journal of the Global South, 6*(1), 77–105.

Arn, J. (1995). Pathway to the periphery: Urbanization, creation of a relative surplus population, and political outcomes in Manila, Philippines. *Urban Anthropology and Studies of Cultural Systems and World Economic Development, 24*(3–4), 189–228.

Asian Development Bank. (2005). *An impact evaluation of the development of Genetically Improved Farmed Tilapia.* Manila: Asian Development Bank.

Asuncion, A., de Guzman, F. G., Luto, F. M., & Sedilla, K. (2019). Manual handling in fish port: An ergonomic assessment on the Porters "Kargadors" in Navotas Fishing Port Complex. In R. S. Goonetilleke and W. Karwowski (Eds.), *International Conference on Applied Human Factors and Ergonomics* (pp. 111–119). Cham: Springer.

Baccini, P., & Brunner, P. H. (2012). *Metabolism of the anthroposphere: Analysis, evaluation, design.* Cambridge, MA: MIT Press.

Bair, J. (2009). *Frontiers of commodity chain research.* Redwood City, CA: Stanford University Press.

Bakker, K. J. (2004) *An uncooperative commodity: Privatizing water in England.* Oxford: Oxford University Press.

Bakker, K., & Bridge, G. (2006). Material worlds? Resource geographies and the matter of nature. *Progress in Human Geography, 30*(1), 5–27.

Bakker, I., & Gill, S. (2003). *Power, production, and social reproduction: Human in/security in the global political economy.* New York: Palgrave Macmillan.

Baluyut, E. A. (1989). A review of the utilization of carps in inland fisheries. In T. Petr (Ed.), *Workshop on the use of cyprinids in the fisheries management of larger inland water bodies of the Indo-Pacific and the fourth session of the Indo-Pacific*

Fishery Commission Working Party of Experts on Inland Fisheries (pp. 100–113). Kathmandu, Nepal: FAO.

Bandayrel, J. (1981, March 26). Experts doubt P100M anti-pollution project. *Bulletin Today*, p. 19.

Bankoff, G. (2003). *Cultures of disaster: Society and natural hazard in the Philippines*. London: Routledge.

Banoub, D., Bridge, G., Bustos, B., Ertör, I., González-Hidalgo, M., & de los Reyes, J. A. (2020). Industrial dynamics on the commodity frontier: Managing time, space and form in mining, tree plantations and intensive aquaculture. *Environment and Planning E: Nature and Space*, *38*(6), 1101–1119.

Barles, S. (2007). Feeding the city: Food consumption and flow of nitrogen, Paris, 1801–1914. *Science of the Total Environment*, *375*(1–3), 48–58.

Barney, K. (2009). Laos and the making of a "relational" resource frontier. *Geographical Journal*, *175*(2), 146–159.

Bartels, L. E., Bruns, A., & Simon, D. (2020). Towards situated analyses of uneven peri-urbanisation: An (urban) political ecology perspective. *Antipode*, *52*(5), 1237–1258.

Barua, M. (2019). Animating capital: Work, commodities, circulation. *Progress in Human Geography*, *43*(4), 650–669.

Basiao, Z. U. (1994). *Tilapia, carp and catfish*. Paper presented at the Seminar-workshop on aquaculture development in Southeast Asia and Prospects for sea-farming and searanching, Iloilo City, August 19–23, 1991.

Bautista, A. M., Carlos, M. H., & San Antonio, A. I. (1988). Hatchery production of *Oreochromis niloticus L.* at different sex rations and stocking densities. *Aquaculture*, *73*, 85–95.

Belton, B., & Bush, S. R. (2014). Beyond net deficits: New priorities for an aquacultural geography. *The Geographical Journal*, *180*(1), 3–14.

Belton, B., Haque, M. M., & Little, D. C. (2012). Does size matter? Reassessing the relationship between aquaculture and poverty in Bangladesh. *Journal of Development Studies*, *48*(7), 904–922.

Belton, B., & Little, D. (2008). The development of aquaculture in central Thailand: Domestic demand versus export-led production. *Journal of Agrarian Change*, *8*(1), 123–143.

Belton, B., & Little, D. C. (2011). Immanent and interventionist inland Asian aquaculture development and its outcomes. *Development Policy Review*, *29*(4), 459–484.

Belton, B., & Thilsted, S. H. (2014). Fisheries in transition: Food and nutrition security implications for the global South. *Global Food Security*, *3*(1), 59–66.

Béné, C., Arthur, R., Norbury, H., Allison, E. H., Beveridge, M., Bush, S., ... & Thilsted, S. H. (2016). Contribution of fisheries and aquaculture to food security and poverty reduction: assessing the current evidence. *World Development*, *79*, 177–196.

Béné, C., Steel, E., Luadia, B. K., & Gordon, A. (2009). Fish as the "bank in the water"—Evidence from chronic-poor communities in Congo. *Food Policy*, *34*(1), 108–118.

Bennett, J. (2010). *Vibrant matter: A political ecology of things*. Durham, NC: Duke University Press.

Benton, T. (1989). Marxism and natural limits: An ecological critique and reconstruction. *New Left Review, 178*(1), 51–86.

Bernal, I. (Director). (1976). *Nunal sa tubig* [Motion picture]. Philippines: Seven Stars Productions.

Bernstein, H. (1996). Agrarian questions then and now. *The Journal of Peasant Studies, 24*(1–2), 22–59.

Bernstein, H. (2010). *Class dynamics of agrarian change*. Sterling, VA: Kumarian Press.

Bernstein, H., & Campling, L. (2006). Commodity studies and commodity fetishism II: "Profits with principles"? *Journal of Agrarian Change, 6*(3), 414–447.

Bestor, T. C. (2004). *Tsukiji: The fish market at the center of the world*. Berkeley: University of California Press.

Beveridge, M. C. M. (1984). *Cage and pen fish farming: Carrying capacity models and environmental impact*. Rome: Food and Agriculture Organization.

BFAR. (1981). *Proposed Kilusang Kabuhayan at Kaunlaran Aqua-marine National Plan*. Quezon City: Bureau of Fisheries and Aquatic Resources, Ministry of Natural Resources, Republic of the Philippines.

BFAR. (2005). *Comprehensive national fisheries industry development plan*. Quezon City: Bureau of Fisheries and Aquatic Resources, Department of Agriculture.

Bigornia, J. (1983a, March 23). Lake authority board should all resign now. *Bulletin Today*, p. 6.

Bigornia, J. (1983b, April 13). More gripes arise over fishpens. *Bulletin Today*, p. 6.

Blanc, E., & Strobl, E. (2016). Assessing the impact of typhoons on rice production in the Philippines. *Journal of Applied Meteorology and Climatology, 55*(4), 993–1007.

Blok, A. (2016). Assembling urban riskscapes: Climate adaptation, scales of change and the politics of expertise in Surat, India. *City, 20*(4), 602–618.

Boyce, J. K. (1993). *The political economy of growth and impoverishment in the Marcos era*. Quezon City: Ateneo de Manila University Press.

Boyd, W., Prudham, W. S., & Schurman, R. A. (2001). Industrial dynamics and the problem of nature. *Society & Natural Resources, 14*(7), 555–570.

Braun, B. (2005). Environmental issues: Writing a more-than-human urban geography. *Progress in Human Geography, 29*(5), 635–650.

Braun, B. (2015). Futures: Imagining socioecological transformation—an introduction. *Annals of the Association of American Geographers, 105*(2), 239–243.

Breitbach, C. (2007). The geographies of a more just food system: Building landscapes for social reproduction. *Landscape Research, 32*(5), 533–557.

Brenner, N., & Schmid, C. (2015). Towards a new epistemology of the urban? *City, 19*(2–3), 151–182.

Bridge, G. (2000). The social regulation of resource access and environmental impact: Production, nature and contradiction in the US copper industry. *Geoforum, 31*(2), 237–256.

Bridge, G. (2001). Resource triumphalism: Postindustrial narratives of primary commodity production. *Environment and Planning A*, *33*(12), 2149–2173.
Buckley, M., & Strauss, K. (2016). With, against and beyond Lefebvre: Planetary urbanization and epistemic plurality. *Environment and Planning D: Society and Space*, *34*(4), 617–636.
Bulletin Today. (1981a, June 21). Fishpen owners sue lake agency, p. 4.
Bulletin Today. (1981b, July 1). LLDA (not Laguna court) foils FM fishpen demolition orders, p. 18.
Bulletin Today. (1981c, September 16). Oppose dismantling of Laguna de Bay fishpens, p. 18.
Bulletin Today. (1982a, March 12). Demolition up as court rules, p. 32.
Bulletin Today. (1982b, April 29). Seek reforms in LL administration, p. 32.
Bulletin Today. (1983a, March 15). Small fishers hit LLDA on fishpens, p. 40.
Bulletin Today. (1983b, April 14). Fishers may get fishpens, pp. 1 & 10.
Bulletin Today. (1983c, December 13). Laguna fish project bared, pp. 1 & 8.
Bulletin Today. (1984a, March 1). LLDA pledges justice to slain fisherman, p. 28.
Bulletin Today. (1984b, March 8). Fishers' protection ordered, p. 1.
Bulletin Today. (1984c, March 9). Drive against armed men in lake ordered, p. 32.
Bulletin Today. (1984d, April 19). Laguna de Bay: problems & plans, p. 1.
Bulletin Today. (1984e, April 10). Coop program in Laguna de Bay, p. 1.
Bunting, S. W., & Little, D. C. (2015). Urban aquaculture for resilient food systems. In H. de Zeeuw & P. Deschel (Eds.), *Cities and agriculture: Developing resilient urban food systems* (pp. 312–335). London: Routledge.
Burgonio, T. (2014, June 21). Aquino OKs P123–B Laguna Lake road dike, 2 other PPP projects. *Philippine Daily Inquirer*. https://newsinfo.inquirer.net/613117/aquino-oks-p123-b-laguna-lake-road-dike-2-other-ppp-projects
BusinessWorld. (2001, November 22). Japan, GEF grants to help LLDA protect Laguna Lake.
Calica, A. (2012, August 14). 195,000 families in danger zones face relocation. *The Philippine Star*. http://www.philstar.com/headlines/2012/08/14/838098/195000-families-danger-zones-face-relocation
Caliwag, F. M. (1966, December 4). A grand project for Laguna Lake. *Sunday Times Magazine*, pp. 26–29.
Calleja, N. (2012, May 23). LLDA: Spread of knife fish an opportunity, not crisis. *Philippine Daily Inquirer*. https://newsinfo.inquirer.net/199329/llda-spread-of-knife-fish-an-opportunity-not-crisis#ixzz31xs5AdXf
Campling, L. (2012). The tuna "commodity frontier": Business strategies and environment in the industrial tuna fisheries of the Western Indian Ocean. *Journal of Agrarian Change*, *12*(2–3), 252–278.
Campling, L., & Havice, E. (2014). The problem of property in industrial fisheries. *Journal of Peasant Studies*, *41*(5), 707–727.
Camus, M. R. (2016, March 28). DPWH declares failed bid for Laguna Lakeshore Expressway Dike. *Philippine Daily Inquirer*. https://business.inquirer.net/208962/dpwh-declares-failed-bid-for-p123-b-laguna-lakeshore-expressway-dike

Canlas, C. (1991). *Calabarzon project: The peasants' scourge.* Philippine Peasant Institute.
Cardenas, M. (1983, April 1). Mayors made scapegoats in fishpen fiasco. *Bulletin Today*, p. 8.
Carlos, M. C. (1994, November 23). "Katring" damage to Laguna Lake at P655.41 million, reports lake dev't authority. *BusinessWorld*, p. 12.
Carlos, M. C. (1995a, September 6). LLDA conflict with LGUs derails phaseout plan for lake's fish pens. *BusinessWorld*, p. 11.
Carlos, M. C. (1995b, November 6). LLDA, LGUs finally agree to lower Laguna Lake fishing area bounds. *BusinessWorld*, p. 10.
Carlos, M. C. (1995c, November 9). Rosing helps DENR in Laguna Lake. *BusinessWorld*, p. 9.
Carnaje, G. P. (2007). *Contractual arrangements in Philippine fisheries* (No. 2007-22). PIDS Discussion Paper Series. Makati: PIDS.
Carse, A. (2012). Nature as infrastructure: Making and managing the Panama Canal watershed. *Social Studies of Science, 42*(4), 539–563.
Castree, N. (1995). The nature of produced nature: Materiality and knowledge construction in Marxism. *Antipode, 27*(1), 12–48.
Castree, N. (2004). The geographical lives of commodities: problems of analysis and critique. *Social & Cultural Geography, 5*(1), 21–35.
Castro, J. M. C., Camacho, M. V. C., & Gonzales, J. C. B. (2018). Reproductive biology of invasive knifefish (*Chitala ornata*) in Laguna de Bay, Philippines and its implication for control and management. *Asian Journal of Conservation Biology, 7*(2), 113–118.
CBBRC, EILER, & WAC. (2011). *Liyab at Alipato: Mga karanasan sa pag-oorganisa sa mga Espesyal na Sonang Pang-ekonomiya sa Pilipinas.* Quezon City: Crispin B. Beltran Resource Center, Inc., Ecumenical Institute for Labor Education and Research and Workers Assistance Center.
Celis, J. M. (1988). Effect of trading time on the quality of fish traded at Navotas Fishing Port Complex. In *Proceedings of the Twentieth Anniversary Seminar on Development of Fish Products in Southeast Asia, Singapore, 27–31 October 1987* (pp. 91–94). Marine Fisheries Research Department, Southeast Asian Fisheries Development Center.
Cendana, S. M., & Mane, A. M. (1937). Recent physical changes in the water of Laguna de Bay and their effect on the lake fauna. *Philippine Agriculturist, 26,* 327–337.
Chavez, H. M., Casao, E. A., Villanueva, E. P., Paras, M. P., Guinto, M. C., & Mosqueda, M. B. (2006a). Heavy metal and microbial analyses of janitor fish (*Pterygoplichthus* spp.) in Laguna de Bay, Philippines. *Journal of Environmental Science and Management, 9*(2), 31–40.
Chavez, J. M., De La Paz, R. M., Manohar, S. K., Pagulayan, R. C., & Vi, J. R. C. (2006b). New Philippine record of South American sailfin catfishes (Pisces: Loricariidae). *Zootaxa, 1109*(1), 57–68.
Chevalier, S. (1998). From woollen carpet to grass carpet: Bridging house and garden in an English suburb. In D. Miller (Ed.), *Material cultures: Why some things matter* (pp. 47–72). London: UCL Press.

Cidell, J. (2018). Assembling and re-assembling Asian carp: The Chicago Area Waterways System as a space of urban politics. In K. Ward, A. E. G. Jonas, B. Miller, & D. Wilson (Eds.), *The Routledge handbook on spaces of urban politics* (pp. 426–438). London: Routledge.

Cinco, M. (2011, September 26). Government sued for P4B for canceled dredging. *Philippine Daily Inquirer*. https://newsinfo.inquirer.net/65549/gov%E2%80%99t-sued-for-p4b-for-canceled-dredging

Cinco, M. (2013, August 22). Aquino eyes "mega-dike" to solve Laguna flooding. *Philippine Daily Inquirer*. https://newsinfo.inquirer.net/471861/aquino-eyes-mega-dike-to-solve-laguna-flooding#ixzz3TymFelnP

Cinco, M. (2014, September 4). Hearty dishes from Laguna's predator fish. *Philippine Daily Inquirer*. http://newsinfo.inquirer.net/635152/hearty-dishes-from-lagunas-predator-fish#ixzz3nnNoSAk9

Cinco, M. (2015, February 19). Predator fish in sausages, dumplings. *Philippine Daily Inquirer*. http://newsinfo.inquirer.net/673844/predator-fish-in-sausages-dumplings#ixzz3nnNNRrd2

Clark, B., & Foster, J. B. (2009). Ecological imperialism and the global metabolic rift: Unequal exchange and the guano/nitrates trade. *International Journal of Comparative Sociology, 50*(3–4), 311–334.

Clausen, R., & Clark, B. (2005). The metabolic rift and marine ecology: An analysis of the ocean crisis within capitalist production. *Organization & Environment, 18*(4), 422–444.

Coe, N. M., Dicken, P., & Hess, M. (2008). Global production networks: realizing the potential. *Journal of Economic Geography, 8*(3), 271–295.

Collins, T. W. (2009). The production of unequal risk in hazardscapes: An explanatory frame applied to disaster at the US–Mexico border. *Geoforum, 40*(4), 589–601.

Collins, T. W. (2010). Marginalization, facilitation, and the production of unequal risk: The 2006 Paso del Norte floods. *Antipode, 42*(2), 258–288.

Connolly, C. (2019). Urban political ecology beyond methodological cityism. *International Journal of Urban and Regional Research, 43*(1), 63–75.

Cons, J., & Eilenberg, M. (2019). *Frontier assemblages: The emergent politics of resource frontiers in Asia*. Hoboken, NJ: John Wiley & Sons.

Cook, I. (2004). Follow the thing: Papaya. *Antipode, 36*(4), 642–664.

Cook, I. [et al.]. (2006). Geographies of food: Following. *Progress in Human Geography, 30*(5), 655–666.

Cooke, S. L. (2016). Anticipating the spread and ecological effects of invasive bigheaded carps (*Hypophthalmichthys* spp.) in North America: A review of modeling and other predictive studies. *Biological Invasions, 18*(2), 315–344.

Cornea, N., Zimmer, A., & Véron, R. (2016). Ponds, power and institutions: The everyday governance of accessing urban water bodies in a small Bengali city. *International Journal of Urban and Regional Research, 40*(2), 395–409.

Cousins, J. J. (2017). Volume control: Stormwater and the politics of urban metabolism. *Geoforum, 85*, 368–380.

Coutard, O. (2008). Placing splintering urbanism: Introduction. *Geoforum*, *39*(6), 1815–1820.
Coutard, O., & Rutherford, J. (2015). *Beyond the networked city: Infrastructure reconfigurations and urban change in the North and South*. London: Routledge.
Cronon, W. (1991). *Nature's metropolis: Chicago and the great West*. New York: W. W. Norton.
Cronon, W. (1996). The trouble with wilderness: Or, getting back to the wrong nature. *Environmental history*, *1*(1), 7–28.
Cruz, J.-A. B. (1982). A dying lake. In B. A. R. Foundation (Ed.), *The troubled waters of Laguna Lake* (pp. 1–9). Metro Manila: BINHI Agricultural Resource Foundation.
Dannhaeuser, N. (1986). Aquaculture and land reform: Incongruous conditions in the Philippines. *Human Organization*, *45*(3), 254.
Das, R. J. (2014). Low-wage capitalism, social difference, and nature-dependent production: A study of the conditions of workers in shrimp aquaculture. *Human Geography*, *7*(1): 7–34.
Davidson, M., & Iveson, K. (2015). Beyond city limits: A conceptual and political defense of "the city" as an anchoring concept for critical urban theory. *City*, *19*(5), 646–664.
Davies, A. (2019). Unwrapping the OXO Cube: Josué de Castro and the intellectual history of metabolism. *Annals of the American Association of Geographers*, *109*(3), 837–856.
Davies, J., Lacanilao, F., & Santiago, A. (1986). *Laguna de Bay: Problems and options*. Manila: Haribon Foundation for the Conservation of Natural Resources.
De Angelis, M. (2007). *The beginning of history: Value struggles and global capital*. London: Pluto Press.
De Janvry, A. (1981). *The agrarian question and reformism in Latin America*. Baltimore, MD: Johns Hopkins University Press.
Dela Cruz, C. R. (1982). *Fishpen and cage culture development project in Laguna de Bay, Philippines*. Rome: Food and Agriculture Organization.
Delgado-Ramos, G. C. (2015). Water and the political ecology of urban metabolism: The case of Mexico City. *Journal of Political Ecology*, *22*(1), 98–114.
Delmendo, M. N., & Gedney, R. H. (1976). Laguna de Bay fish pen aquaculture development—Philippines. *Proceedings of the Annual Meeting—World Mariculture Society*, *7*(1–4), 257–265.
Delmendo, M. N., & Rabanal, H. R. (1982). The organization and administration of aquaculture development in Asian countries—part 2. *Agricultural Administration*, *9*, 131–137.
Delos Reyes, M. (1993). Fishpen culture and its impact on the ecosystem of Laguna de Bay, Philippines. In V. Christensen & D. Pauly (Eds.), *Trophic models of aquatic ecosystems* (pp. 74–84). Manila: ICLARM.
Demaria, F., & Schindler, S. (2016). Contesting urban metabolism: Struggles over waste-to-energy in Delhi, India. *Antipode*, *48*(2), 293–313.

Dey, M. M., & Ahmed, M. (2005). Aquaculture—food and livelihoods for the poor in Asia: A brief overview of the issues. *Aquaculture Economics & Management, 9*(1-2), 1-10.

Dinglasan, R. R. (2012, August 14). DPWH: Floodway through congested Paranaque not "practical". *GMA News.* https://www.gmanetwork.com/news/news/nation/269577/dpwh-floodway-through-congested-paranaque-not-practical/story/

Doeppers, D. F. (2016). *Feeding Manila in peace and war, 1850-1945.* Madison: University of Wisconsin Press.

Dooling, S., & Simon, G. (2012). *Cities, nature and development: The politics and production of urban vulnerabilities.* Surrey: Ashgate.

Doshi, S. (2017). Embodied urban political ecology: Five propositions. *Area, 49*(1), 125-128.

DPWTC. (1972). *Manila and suburbs flood control and drainage project: Final project report.* Manila: Department of Public Works and Communications.

Dressler, W. H., & Guieb, E. R., III. (2015). Violent enclosures, violated livelihoods: Environmental and military territoriality in a Philippine frontier. *Journal of Peasant Studies, 42*(2), 323-345.

Eakin, H., & Appendini, K. (2008). Livelihood change, farming, and managing flood risk in the Lerma Valley, Mexico. *Agriculture and Human Values, 25*(4), 555.

Eakin, H., Lerner, A. M., & Murtinho, F. (2010). Adaptive capacity in evolving peri-urban spaces: Responses to flood risk in the Upper Lerma River Valley, Mexico. *Global Environmental Change, 20*(1), 14-22.

Eaton, E. (2011). On the farm and in the field: The production of nature meets the agrarian question. *New Political Economy, 16*(2), 247-251.

Eilenberg, M. (2014). Frontier constellations: Agrarian expansion and sovereignty on the Indonesian-Malaysian border. *Journal of Peasant Studies, 41*(2), 157-182.

Ekers, M., & Loftus, A. (2013). Revitalizing the production of nature thesis: A Gramscian turn? *Progress in Human Geography, 37*(2), 234-252.

Eknath, A. E., & Acosta, B. O. (1998). *Genetic Improvement of Farmed Tilapias (GIFT) Project.* Manila: ICLARM.

Eleazar, F. C. (1992). *Managing common property resources: A case study of Laguna Lake* (Unpublished thesis), University of the Philippines, Quezon City.

Ertör, I., & Ortega-Cerdà, M. (2015). Political lessons from early warnings: Marine finfish aquaculture conflicts in Europe. *Marine Policy, 51*, 202-210.

Ertör, I., & Ortega-Cerdà, M. (2019). The expansion of intensive marine aquaculture in Turkey: The next-to-last commodity frontier? *Journal of Agrarian Change, 19*(2), 337-360.

Esplanada, J. E. (2012, August 2). MWSS: Government needs new water sources. *Philippine Daily Inquirer.* http://newsinfo.inquirer.net/241127/mwss-government-needs-new-water-sources

Etkin, D. (1999). Risk transference and related trends: Driving forces towards more mega-disasters. *Global Environmental Change Part B: Environmental Hazards, 1*(2), 69-75.

Everts, J. (2015). Invasive life, communities of practice, and communities of fate. *Geografiska Annaler: Series B, Human Geography, 97*(2), 195–208.

Faier, L., & Rofel, L. (2014). Ethnographies of encounter. *Annual Review of Anthropology, 43,* 363–377.

FAO. (2006). *State of world aquaculture.* Rome: Food and Agriculture Organization—Fisheries and Aquaculture Department.

FAO. (2018). *The State of world fisheries and aquaculture 2018—Meeting the sustainable development goals.* Rome: Food and Agriculture Organization.

Fermin, A. C. (1991). LHRHa and domperidone-induced oocyte maturation and ovulation in bighead carp. *Aristichthys nobilis* (Richardson). *Aquaculture, 93,* 87–94.

Fernandez, D. G., Mabesa, R. C., & Mabesa, L. B. (1998). Characterization of bighead carp (*Aristichthys nobilis,* Richardson) muscle for surimi production. *UPV Journal of Natural Science, 3*(2), 140–149.

Ferrão, P., & Fernández, J. E. (2013). *Sustainable urban metabolism.* Cambridge, MA: MIT Press.

FIAN-Philippines. (2009). *Right to food condition of urban poor families in Navotas City: A case study.* Quezon City: Foodfirst Information and Action Network Philippines.

Fischer-Kowalski, M. (1998). Society's metabolism: the intellectual history of materials flow analysis, Part I, 1860–1970. *Journal of Industrial Ecology, 2*(1), 61–78.

FitzSimmons, M. (1989). The matter of nature. *Antipode, 21,* 106–120.

Florendo, A. C. (1969, July 5). The grand vision for Laguna Lake. *Mirror,* p. 10.

Fold, N., & Hirsch, P. (2009). Re-thinking frontiers in Southeast Asia. *The Geographical Journal, 175*(2), 95–97.

Foster, J. B. (1999). Marx's theory of metabolic rift: Classical foundations for environmental sociology. *American Journal of Sociology, 105*(2), 366–405.

Franco, J. C., & Borras, S. M., Jr. (2007). Struggles over land resources in the Philippines. *Peace Review: A Journal of Social Justice, 19*(1), 67–75.

Frawley, J., & McCalman, I. (Eds.). (2014). *Rethinking invasion ecologies from the environmental humanities.* London: Routledge.

Freidberg, S. E. (2001a). Gardening on the edge: The social conditions of unsustainability on an African urban periphery. *Annals of the Association of American Geographers, 91*(2), 349–369.

Freidberg, S. E. (2001b). On the trail of the global green bean: Methodological considerations in multi-site ethnography. *Global Networks, 1*(4), 353–368.

Freidberg, S. E. (2004). *French beans and food scares: Culture and commerce in an anxious age.* Oxford: Oxford University Press.

Freidberg, S. (2009). *Fresh.* Cambridge, MA: Harvard University Press.

Furlong, K., & Kooy, M. (2017). Worlding water supply: Thinking beyond the network in Jakarta. *International Journal of Urban and Regional Research, 41*(6), 888–903.

Gabriel, N. (2014). Urban political ecology: Environmental imaginary, governance, and the non-human. *Geography Compass, 8*(1), 38–48.

Gamil, J. T. (2016, August 4). Laguna fish pens must go, Gina ordered. *Philippine Daily Inquirer.* http://newsinfo.inquirer.net/802717/laguna-fish-pens-must-go-gina-ordered#ixzz4t8IZOgvH

Gandy, M. (2003). *Concrete and clay: Reworking nature in New York City.* Cambridge, MA: MIT Press.

Gandy, M. (2004). Rethinking urban metabolism: Water, space and the modern city. *City, 8*(3), 363–379.

Gandy, M. (2005). Cyborg urbanization: complexity and monstrosity in the contemporary city. *International Journal of Urban and Regional Research, 29*(1), 26–49.

Gandy, M. (2006). Planning, anti-planning and the infrastructure crisis facing metropolitan Lagos. *Urban Studies, 43*(2), 371–396.

Garcia, A. M., & Medina, R. T. (1987). The state of development program of cage culture in Laguna Lake. In L. C. Davin, M. A. Mangaser, & Z. C. Gibe (Eds.), *State of development of the Laguna de Bay area* (pp. 17–24). Los Banos: PCARRD.

Garcia, Y. T., Dey, M. M., & Navarez, S. M. M. (2005). Demand for fish in the Philippines: A disaggregated analysis. *Aquaculture Economics & Management, 9*(1), 141–168.

Garrido, M. Z. (2019). *The patchwork city: Class, space, and politics in Metro Manila.* Chicago: University of Chicago Press.

Gereffi, G., Korzeniewicz, M., & Korzeniewicz, R. P. (1994). Introduction: Global commodity chains. In G. Gereffi & M. Korzeniewicz (Eds.), *Commodity chains and global capitalism* (pp. 1–14). Westport, CT: Greenwood Publishing Group.

Ghosh, S., & Meer, A. (2021). Extended urbanisation and the agrarian question: Convergences, divergences and openings. *Urban Studies, 58*(6), 1097–1119.

Glassman, J. (2006). Primitive accumulation, accumulation by dispossession, accumulation by "extra-economic" means. *Progress in Human Geography, 30*(5), 608–625.

Goldoftas, B. (2006). *The green tiger: The costs of ecological decline in the Philippines.* Oxford: Oxford University Press.

Goldstein, B., Birkved, M., Fernández, J., & Hauschild, M. (2017). Surveying the environmental footprint of urban food consumption. *Journal of Industrial Ecology, 21*(1), 151–165.

Golubiewski, N. (2012). Is there a metabolism of an urban ecosystem? An ecological critique. *Ambio, 41*(7), 751–764.

Gonzales, E. R. (1984). Small scale tilapia cage technology adopted in fishing villages in Laguna Lake, Philippines. *Aquaculture, 41*, 161–169.

Goodman, D., Sorj, B., & Wilkinson, J. (1987). *From farming to biotechnology: A theory of agro-industrial development.* Oxford: Basil Blackwell.

Goss, J., Skladany, M., & Middendorf, G. (2001). Dialogue: Shrimp aquaculture in Thailand: A response to Vandergeest, Flaherty, and Miller. *Rural Sociology, 66*(3), 451.

Graham, S. (Ed.). (2010). *Disrupted cities: When infrastructure fails.* London: Routledge.

Graham, S., & Marvin, S. (2001). *Splintering urbanism: Networked infrastructures, technological mobilities and the urban condition.* London: Routledge.

Graham, S., & McFarlane, C. (2014). *Infrastructural lives: Urban infrastructure in context*. London: Routledge.

Graham, S., & Thrift, N. (2007). Out of order: Understanding repair and maintenance. *Theory, Culture & Society, 24*(3), 1–25.

Green, S. J., White, A. T., Flores, J. O., Carreon, M. F. I., & Sia, A. E. (2003). *Philippine fisheries in crisis: A framework for management*. Cebu City: Coastal Resource Management Project of the Department of Environment and Natural Resources.

Guerrero, R. D., III (1981). *Introduction to fish culture in the Philippines*. Manila: Philippine Education Co.

Guerrero, R. D., III (2014). Impacts of introduced freshwater fishes in the Philippines (1905–2013): A review and recommendations. *Philippine Journal of Science, 143*(1): 49–59.

Guibrunet, L., Calvet, M. S., & Broto, V. C. (2017). Flows, system boundaries and the politics of urban metabolism: Waste management in Mexico City and Santiago de Chile. *Geoforum, 85*, 353–367.

Gururani, S. (2020). Cities in a world of villages: Agrarian urbanism and the making of India's urbanizing frontiers. *Urban Geography, 41*(7), 971–989.

Gustafson, S. (2015). The making of a landslide: Legibility and expertise in exurban southern Appalachia. *Environment and Planning A: Economy and Space, 47*(7), 1404–1421.

Gustafson, S., Heynen, N., Rice, J. L., Gragson, T., Shepherd, J. M., & Strother, C. (2014). Megapolitan political ecology and urban metabolism in Southern Appalachia. *The Professional Geographer, 66*(4), 664–675.

Guthman, J. (2011). Bodies and accumulation: Revisiting labour in the "Production of Nature". *New Political Economy, 16*(2), 233–238.

Guzman, R. D., Torres, R. D., & Darrah, L. B. (1974). *The impact of bangus landing from Laguna Lake (Rizal points) on bangus prices in Malabon*. Quezon City: National Food and Agriculture Council.

Hall, D. (2004). Explaining the diversity of Southeast Asian shrimp aquaculture. *Journal of Agrarian Change, 4*(3), 315–335.

Hamilton-Hart, N., & Stringer, C. (2016). Upgrading and exploitation in the fishing industry: Contributions of value chain analysis. *Marine Policy, 63*, 166–171.

Harvey, D. (1996). *Justice, nature and the geography of difference*. Cambridge: Blackwell.

Harvey, D. (2003). *The new imperialism*. Oxford: Oxford University Press.

Harvey, D. (2006). *The limits to capital*. London: Verso.

Havice, E., & Reed, K. (2012). Fishing for development? Tuna resource access and industrial change in Papua New Guinea. *Journal of Agrarian Change, 12*(2–3), 413–435.

Head, L., & Atchison, J. (2015). Entangled invasive lives: Indigenous invasive plant management in northern Australia. *Geografiska Annaler: Series B, Human Geography, 97*(2), 169–182.

Heistermann, M., Crisologo, I., Abon, C. C., Racoma, B. A., Jacobi, S., Servando, N. T., ... & Bronstert, A. (2013). Using the new Philippine radar network to reconstruct the Habagat of August 2012 monsoon event around Metropolitan Manila. *Natural Hazards & Earth System Sciences, 13*(3).

Henderson, G. L. (1999). *California and the fictions of capital*. Oxford: Oxford University Press.

Herre, A. W., & Mendoza, J. (1929). Bangos culture in the Philippine Islands. *Philippine Journal of Science, 38*(4), 451–509.

Heynen, N. (2014). Urban political ecology I: The urban century. *Progress in Human Geography, 38*(4), 598–604.

Heynen, N., Kaika, M., & Swyngedouw, E. (2006). *In the nature of cities: Urban political ecology and the politics of urban metabolism*. London: Routledge.

Hinkle, H. B. (1950). *A report on the transportation and marketing of fresh fish in the Manila area*. Unpublished report.

Hodkinson, S. (2012). The new urban enclosures. *City, 16*(5), 500–518.

Holifield, R. (2009). Actor-network theory as a critical approach to environmental justice: A case against synthesis with urban political ecology. *Antipode, 41*(4), 637–658.

Hommes, L., & Boelens, R. (2017). Urbanizing rural waters: Rural-urban water transfers and the reconfiguration of hydrosocial territories in Lima. *Political Geography, 57*, 71–80.

Howe, C., Lockrem, J., Appel, H., Hackett, E., Boyer, D., Hall, R., . . . & Ballestero, A. (2016). Paradoxical infrastructures: Ruins, retrofit, and risk. *Science, Technology, & Human Values, 41*(3), 547–565.

Huang, S. L. (1998). Urban ecosystems, energetic hierarchies, and ecological economics of Taipei metropolis. *Journal of Environmental Management, 52*(1), 39–51.

Huber, M. T. (2017). Hidden abodes: Industrializing political ecology. *Annals of the American Association of Geographers, 107*(1), 151–166.

Huber, M. T., & Emel, J. (2009). Fixed minerals, scalar politics: The weight of scale in conflicts over the "1872 Mining Law" in the United States. *Environment and Planning A, 41*(2), 371–388.

Hudson, R. (2008). Cultural political economy meets global production networks: A productive meeting? *Journal of Economic Geography, 8*(3), 421–440.

Hughes, A., & Reimer, S. (Eds.). (2004). *Geographies of commodity chains*. London: Routledge.

Israel, D. C. (2007). *The current state of aquaculture in Laguna de Bay*. Makati City: Philippine Institute of Development Studies.

Israel, D. C., Boni-Cortez, M. C., & Patambang, M. E. (2008). *Aquaculture development in Laguna de Bay: An economic analysis*. Tigbauan and Makati City: Southeast Asian Fisheries Development Center—Aquaculture Department and Philippine Institute of Development Studies.

Ito, S. (2002). From rice to prawns: Economic transformation and agrarian structure in rural Bangladesh. *The Journal of Peasant Studies, 29*(2), 47–70.

Jazeel, T. (2018). Urban theory with an outside. *Environment and Planning D: Society and Space, 36*(3), 405–419.

Jepson, W., Brannstrom, C., & Filippi, A. (2010). Access regimes and regional land change in the Brazilian Cerrado, 1972–2002. *Annals of the Association of American Geographers, 100*(1), 87–111.

Jha, A. K., Bloch, R., & Lamond, J. (2012). *Cities and flooding: A guide to integrated urban flood risk management for the 21st century.* Washington, DC: The World Bank.

Jose, D. (1994a, March 6). Demolition of illegal fishpens halted: Part 1. *Philippine Daily Inquirer*, pp. 1, 12.

Jose, D. (1994b, March 7). Demolition of illegal fishpens halted: Part 2. *Philippine Daily Inquirer*, pp. 1, 11.

Kaika, M. (2004). *City of flows: Modernity, nature, and the city.* London: Routledge.

Kaika, M., & Swyngedouw, E. (2000). Fetishizing the modern city: The phantasmagoria of urban technological networks. *International Journal of Urban and Regional Research, 24*(1), 120–138.

Kanai, J. M. (2014). On the peripheries of planetary urbanization: Globalizing Manaus and its expanding impact. *Environment and Planning D: Society and Space, 32*(6), 1071–1087.

Kanai, J. M., & Schindler, S. (2019). Peri-urban promises of connectivity: Linking project-led polycentrism to the infrastructure scramble. *Environment and Planning A: Economy and Space, 51*(2), 302–322.

Kaplinsky, R. (2000). Globalization and unequalization: What can be learned from value chain analysis?. *Journal of Development Studies, 37*(2), 117–146.

Karaos, A. M. A. (1993). Manila's squatter movement: A struggle for place and identity. *Philippine Sociological Review, 41*(1/4), 71–91.

Karvonen, A. (2011). *Politics of urban runoff: Nature, technology, and the sustainable city.* Cambridge, MA: MIT Press.

Katz, C. (2001). Vagabond capitalism and the necessity of social reproduction. *Antipode, 33*(4), 709–728.

Kautsky, K. (1988). *The agrarian question.* Winchester, MA: Zwan Publications.

Keil, R., & Macdonald, S. (2016). Rethinking urban political ecology from the outside in: Greenbelts and boundaries in the post-suburban city. *Local Environment, 21*(12), 1516–1533.

Kelly, P. F. (1996). Blue revolution or red herring? Fish farming and development discourse in the Philippines. *Asia Pacific Viewpoint, 37*(1), 39–58.

Kelly, P. F. (2000). *Landscapes of globalization: Human geographies of economic change in the Philippines.* London: Routledge.

Kelly, P. F. (2013). Production networks, place and development: Thinking through global production networks in Cavite, Philippines. *Geoforum, 44*, 82–92.

Kennedy, C., Cuddihy, J., & Engel-Yan, J. (2007). The changing metabolism of cities. *Journal of Industrial Ecology, 11*(2), 43–59.

Kerkvliet, B. J. (2002). *The Huk rebellion: A study of peasant revolt in the Philippines.* Lanham, MD: Rowman & Littlefield.

Khaw, H. L., Ponzoni, R. W., & Danting, M. J. C. (2008). Estimation of genetic change in the GIFT strain of Nile tilapia (Oreochromis niloticus) by comparing contemporary progeny produced by males born in 1991 or in 2003. *Aquaculture*, *275*(1–4), 64–69.

Kleibert, J. M., & Kippers, L. (2016). Living the good life? The rise of urban mixed-use enclaves in Metro Manila. *Urban Geography*, *37*(3), 373–395.

Kloppenburg, J. R. (2005). *First the seed: The political economy of plant biotechnology*. Madison: University of Wisconsin Press.

Kolar, C. S., Chapman, D. C., Courtenay, W. R. J., Housel, C. M., Williams, J. D., & Jennings, D. P. (2005). *Asian carps of the Genus Hypophthalmicthys (Pisces, Cyprinidae)—a biological synopsis and environmental risk assessment*. Washington, DC: US Fish and Wildlife Service.

Kooy, M., & Bakker, K. (2008). Technologies of government: Constituting subjectivities, spaces, and infrastructures in colonial and contemporary Jakarta. *International Journal of Urban and Regional Research*, *32*(2), 375–391.

Kopytoff, I. (1986). The cultural biography of things: Commoditization as process. In A. Appadurai (Ed.), *The social life of things: Commodities in cultural perspective* (pp. 64–92). Cambridge: Cambridge University Press.

Krause, G., Brugere, C., Diedrich, A., Ebeling, M. W., Ferse, S. C., Mikkelsen, E., . . . & Troell, M. (2015). A revolution without people? Closing the people–policy gap in aquaculture development. *Aquaculture*, *447*, 44–55.

Laguna Lake Development Authority. (1966). *The Laguna Lake Development Authority prospectus*. Manila: Laguna Lake Development Authority.

Laguna Lake Development Authority. (1970). *Profile of fishery development with a concise project study on Looc Pilot Lake Fishing*. Pasig: Laguna Lake Development Authority.

Laguna Lake Development Authority. (1971). *General manager's second annual report 1971*. Pasig: Laguna Lake Development Authority.

Laguna Lake Development Authority. (1972). *General manager's third annual report 1972*. Pasig: Laguna Lake Development Authority.

Laguna Lake Development Authority. (1974). *LLDA annual financial report calendar year 1974*. Pasig: Laguna Lake Development Authority.

Laguna Lake Development Authority. (1977). *Annual report 1977*. Pasig: Laguna Lake Development Authority.

Laguna Lake Development Authority. (1978a). *Annual report 1978*. Pasig: Laguna Lake Development Authority.

Laguna Lake Development Authority. (1978b). *Final report—Comprehensive Water Quality Management Program, Laguna de Bay*. Pasig: Laguna Lake Development Authority.

Laguna Lake Development Authority. (1979). *Annual report 1979*. Pasig: Laguna Lake Development Authority.

Laguna Lake Development Authority. (1980). *Annual report 1980*. Pasig: Laguna Lake Development Authority.

Laguna Lake Development Authority. (1981). *Annual report 1981*. Pasig: Laguna Lake Development Authority.

Laguna Lake Development Authority. (1982). *Laguna Lake Cooperative Development Program*. Pasig: Laguna Lake Development Authority.

Laguna Lake Development Authority. (1983). *Laguna Lake Fishery Zoning and Management Plan*. Pasig: Laguna Lake Development Authority.

Laguna Lake Development Authority. (1985). *Annual report 1985*. Pasig: Laguna Lake Development Authority.

Laguna Lake Development Authority. (1986). *Annual report 1986*. Pasig: Laguna Lake Development Authority.

Laguna Lake Development Authority. (1995a). *Annual report 1995*. Pasig: Laguna Lake Development Authority.

Laguna Lake Development Authority. (1995b). *The Laguna de Bay master plan final report*. Pasig: Laguna Lake Development Authority.

Laguna Lake Development Authority. (1999). *Celebrating 30 years of lake management*. Taytay, Rizal: Laguna Lake Development Authority/Department of Environment and Natural Resources.

Lara, F., Jr., & Morales, H. R., Jr. (1990). The peasant movement and the challenge of rural democratisation in the Philippines. *The Journal of Development Studies, 26*(4), 143–162.

Larkin, B. (2013). The politics and poetics of infrastructure. *Annual Review of Anthropology, 42*, 327–343.

Lasco, R. D., & Espaldon, M. V. O. (Eds.). (2005). *Ecosystems and people: The Philippine millennium ecosystem assessment (MA) sub-global assessment*. College, Laguna: Environmental Forestry Programme, College of Forestry and Natural Resources, University of the Philippines Los Banos.

Lawhon, M. (2013). Flows, friction and the sociomaterial metabolization of alcohol. *Antipode, 45*(3), 681–701.

Lawhon, M., Ernstson, H., & Silver, J. (2014). Provincializing urban political ecology: Towards a situated UPE through African urbanism. *Antipode, 46*(2), 497–516.

Lebel, L., Sinh, B. T., Garden, P., Hien, B. V., Subsin, N., Tuan, L., & Vinh, N. (2009). Risk reduction or redistribution? Flood management in the Mekong region. *Asian Journal of Environment and Disaster Management, 1*(1), 23–39.

Lepawsky, J., Akese, G., Billah, M., Conolly, C., & McNabb, C. (2015). Composing urban orders from rubbish electronics: Cityness and the site multiple. *International Journal of Urban and Regional Research, 39*(2), 185–199.

Li, T. M. (2007). Practices of assemblage and community forest management. *Economy and Society, 36*(2), 263–293.

Li, T. M. (2010). To make live or let die? Rural dispossession and the protection of surplus populations. *Antipode, 41*, 66–93.

Li, T. M. (2014). *Land's end: Capitalist relations on an indigenous frontier*. Durham, NC: Duke University Press.

Lico, G. (2003). *Edifice complex: Power, myth, and Marcos state architecture*. Quezon City: Ateneo University Press.

Liongson, L. Q. (2008). Flood mitigation in Metro Manila. *Philippine Engineering Journal, 29*(1), 51–66.

Little, D. C., & Bunting, S. W. (2005). Opportunities and constraints to urban aquaculture, with a focus on South and Southeast Asia. In B. Costa-Pierce, A. Desbonnet, P. Edwards, & D. Baker (Eds.), *Urban aquaculture* (pp. 25–44). Oxfordshire, OX: CABI Publishing.

Loftus, A. (2006). Reification and the dictatorship of the water meter. *Antipode, 38*(5), 1023–1045.

Loftus, A. (2012). *Everyday environmentalism: Creating an urban political ecology*. Minneapolis: University of Minnesota Press.

Loftus, A., & March, H. (2016). Financializing desalination: Rethinking the returns of big infrastructure. *International Journal of Urban and Regional Research, 40*(1), 46–61.

Longo, S. B., Clausen, R., & Clark, B. (2015). *The tragedy of the commodity: Oceans, fisheries, and aquaculture*. New Brunswick, NJ: Rutgers University Press.

Lumbera, B. (2010). Philippine theater in confinement: Breaking out of martial law. *Kritika Kultura, 14*, 97–102.

Luyt, B. (1995). The politics of JICA in the Philippines. *Journal of Contemporary Asia, 25*(3), 380–396.

Macatuno, E. (1966, December 4). A grand project for Laguna Lake. *Sunday Times Magazine*, pp. 26–29.

Magno, F. A. (1993). Politics, elites and transformation in Malabon. *Philippine Studies, 41*(2), 204–216.

Mane, E. C. (1983, September 7). Here's how the fishpen controversy started. *Bulletin Today*, p. 31.

Mane, A. M., & Villaluz, D. K. (1939). The pukot fisheries of Laguna de Bay. *Philippine Journal of Science, 69*(4), 397–414.

Manila Municipal Board. (1905a). *Report of the Municipal Board of the City of Manila for the fiscal year ended June 30, 1904*. Manila: Bureau of Public Printing.

Manila Municipal Board. (1905b). *Report of the Municipal Board of the City of Manila for the fiscal year ended June 30, 1905*. Manila: Bureau of Printing.

Manila Municipal Board. (1910). *Annual report of the Municipal Board of the City of Manila for the fiscal year 1910*. Manila: Bureau of Printing.

Manila Standard Today. (2012, July 31). Agency to develop new water sources,. http://manilastandardtoday.com/www2/2012/07/31/agency-to-develop-new-water-sources/

Manila Water. (2011). *2011 annual report*. Quezon City: Manila Water Company, Inc.

Manipol, L. M. (1981a, January 15). Illegal Laguna de Bay fishpens' demolition set. *Bulletin Today*, p. 5.

Manipol, L. M. (1981b, March 3). Fishpen owners defy order. *Bulletin Today*, p. 18.

Mann, S. (1990). *Agrarian capitalism in theory and practice*. Chapel Hill, NC: UNC Press Books.

Mann, S. A., & Dickinson, J. M. (1978). Obstacles to the development of a capitalist agriculture. *The Journal of Peasant Studies*, *5*(4), 466–481.

Mansfield, B. (2004). Rules of privatization: Contradictions in neoliberal regulation of North Pacific fisheries. *Annals of the Association of American Geographers*, *94*(3), 565–584.

Mansfield, B. (2011). Is fish health food or poison? Farmed fish and the material production of un/healthy nature. *Antipode*, *43*(2), 413–434.

March, H. (2015). Taming, controlling and metabolizing flows: Water and the urbanization process of Barcelona and Madrid (1850–2012). *European Urban and Regional Studies*, *22*(4), 350–367.

Marcos, F. E. (1983). Executive order no. 927. Manila, Philippines. http://www.official gazette.gov.ph/1983/12/16/executive-order- no-927-s-1983/

Marks, D. (2019). Assembling the 2011 Thailand floods: Protecting farmers and inundating high-value industrial estates in a fragmented hydro-social territory. *Political Geography*, *68*, 66–76.

Marvin, S., & Medd, W. (2006). Metabolisms of obe-city: Flows of fat through bodies, cities, and sewers. *Environment and Planning A*, *38*(2), 313–324.

Massey, D. (1994). *Space, place and gender*. Minneapolis: University of Minnesota Press.

Massey, D. (2005). *For space*. London: Sage.

Maynilad. (2010). *2010 Annual report: Witnessing progress*. Quezon City: Maynilad Water Services, Inc.

Mayuga, J. L. (2012, August 19). Rizal, Laguna asked to rethink Laguna Lake spillway. *Business Mirror*. http://businessmirror.com.ph/home/regions/31479-rizal -laguna-asked-to-rethink-laguna-lake-spillway

McCoy, A. W. (2017). Global populism: A lineage of Filipino strongmen from Quezon to Marcos and Duterte. *Kasarinlan: Philippine Journal of Third World Studies*, *32*(1, 2), 7–54.

McFarlane, C. (2011). The city as assemblage: Dwelling and urban space. *Environment and Planning D: Society and Space*, *29*(4), 649–671.

McFarlane, C. (2013). Metabolic inequalities in Mumbai: Beyond telescopic urbanism. *City*, *17*(4), 498–503.

McFarlane, C., Desai, R., & Graham, S. (2014). Informal urban sanitation: Everyday life, poverty, and comparison. *Annals of the Association of American Geographers*, *104*(5), 989–1011.

McFarlane, C., & Rutherford, J. (2008). Political infrastructures: Governing and experiencing the fabric of the city. *International Journal of Urban and Regional Research*, *32*(2), 363–374.

McFarlane, C., & Silver, J. (2017). The political city: "Seeing sanitation" and making the urban political in Cape Town. *Antipode*, *49*(1), 125–148.

McGregor, J., & Chatiza, K. (2019). Frontiers of urban control: Lawlessness on the city edge and forms of clientalist statecraft in Zimbabwe. *Antipode*, *51*(5), 1554–1580.

McKinnon, I., Hurley, P. T., Myles, C. C., Maccaroni, M., & Filan, T. (2019). Uneven urban metabolisms: Toward an integrative (ex) urban political ecology of sustainability in and around the city. *Urban Geography, 40*(3), 352–377.

McMichael, P. (2003). Food security and social reproduction: Issues and contradictions. In I. Bakker & S. Gill (Eds.), *Power, production and social reproduction* (pp. 169–189). London: Palgrave Macmillan.

Mehzabeen, E. (2019). Hinterlands. In A. Orum (Ed.), *The Wiley Blackwell encyclopedia of urban and regional studies* (pp. 816–821). Malden, MA: Wiley-Blackwell.

Melosi, M. V. (2008). *The sanitary city: Environmental services in urban America from colonial times to the present.* Pittsburgh: University of Pittsburgh Press.

Mercene, E. C. (1987). Assessment of capture fisheries and biology of some commercially important fish species. In L. C. Darvin, M. A. Mangaser, & Z. C. Gibe (Eds.), *State of development of the Laguna de Bay area* (pp. 10–16). Los Banos: PCARRD.

Mijares, P. (2017). *The conjugal dictatorship of Ferdinand and Imelda Marcos: Revised and annotated edition.* Quezon City: Ateneo de Manila University Press.

Miles, M. B., & Huberman, A. M. (1994). *Qualitative data analysis: An expanded sourcebook.* Thousand Oaks, CA: Sage.

Miller, D. (2005). *Materiality.* Durham, NC: Duke University Press.

Mitchell, K., Marston, S. A., & Katz, C. (2004). Life's work: An introduction, review and critique. In K. Mitchell, S. A. Marston, & C. Katz (Eds.), *Life's work: Geographies of social reproduction* (pp. 1–26). Malden, MA: Wiley.

Mitchell, T. (2002). *Rule of experts: Egypt, techno-politics, modernity.* Berkeley: University of California Press.

Mohan, S. (2016). Institutional change in value chains: Evidence from tea in Nepal. *World Development, 78,* 52–65.

Molina, V. B. (2012). Non-carcinogenic health risks of heavy metal in mudfish from Laguna Lake. *Science Diliman, 24*(1).

Molina, V. B., Espaldon, M. V. O., Flavier, M. E., Pacardo, E. P., & Rebancos, C. M. (2011). Bioaccumulation in Nile Tilapia (Oreochromis niloticus) from Laguna de Bay, Philippines. *Journal of Environmental Science and Management, 14*(2), 28–35.

Monstadt, J. (2009). Conceptualizing the political ecology of urban infrastructures: Insights from technology and urban studies. *Environment and Planning A, 41*(8), 1924–1942.

Monstadt, J., & Schramm, S. (2017). Toward the networked city? Translating technological ideals and planning models in water and sanitation systems in Dar es Salaam. *International Journal of Urban and Regional Research, 41*(1), 104–125.

Moore, J. W. (2011). Transcending the metabolic rift: A theory of crises in the capitalist world-ecology. *The Journal of Peasant Studies, 38*(1), 1–46.

Moore, J. W. (2015). *Capitalism in the web of life: Ecology and the accumulation of capital.* London: Verso Books.

Morgan, G. R., & Staples, D. J. (2006). *The history of industrial marine fisheries in Southeast Asia.* Bangkok: Food and Agriculture Organization.

Morley, I. (2016). Modern urban designing in the Philippines, 1898–1916. *Philippine Studies Historical & Ethnographic Viewpoints, 64*(1), 3–42.
Morley, I. (2018). Manila. *Cities, 72*, 17–33.
Mouton, M., & Shatkin, G. (2020). Strategizing the for-profit city: The state, developers, and urban production in Mega Manila. *Environment and Planning A: Economy and Space, 52*(2), 403–422.
MPWTC (1952). *Plan for the drainage of Manila and suburbs*. Manila: Bureau of Public Works.
Mustafa, D. (2005). The production of an urban hazardscape in Pakistan: Modernity, vulnerability, and the range of choice. *Annals of the Association of American Geographers, 95*(3), 566–586.
Myers, G. A. (2008). Peri-urban land reform, political-economic reform, and urban political ecology in Zanzibar. *Urban Geography, 29*(3), 264–288.
Napoletano, B. M., Foster, J. B., Clark, B., Urquijo, P. S., McCall, M. K., & Paneque-Gálvez, J. (2019). Making space in critical environmental geography for the metabolic rift. *Annals of the American Association of Geographers, 109*(6), 1811–1828.
Natale, F., Hofherr, J., Fiore, G., & Virtanen, J. (2013). Interactions between aquaculture and fisheries. *Marine Policy, 38*, 205–213.
National Statistical Coordinating Board. (1999). *Estimation of fish biomass in Laguna de Bay based on primary productivity*. Makati: National Statistical Coordinating Board.
Natividad, B. T. (2005, October 7). Saving the Laguna Lake. *BusinessWorld*, p. 6.
Navera, E. (1976). *Fish marketing at the Navotas Fish Landing and Market Authority* (Unpublished thesis), University of the Philippines at Los Banos, College, Laguna.
Navotas. (2010). *Socio economic profile*. Navotas: City of Navotas.
Nayak, P. K., & Berkes, F. (2010). Whose marginalisation? Politics around environmental injustices in India's Chilika lagoon. *Local Environment, 15*(6), 553–567.
Nevins, J., & Peluso, N. L. (Eds.). (2008). *Taking Southeast Asia to market: Commodities, nature, and people in the neoliberal age*. Ithaca, NY: Cornell University Press.
Newell, J. P., & Cousins, J. J. (2015). The boundaries of urban metabolism: Towards a political–industrial ecology. *Progress in Human Geography, 39*(6), 702–728.
Ng, W. (1983, March 30). KKK may run fishpens. *Bulletin Today*, pp. 1 & 16.
Nilo, P., & Espinueva, S. (2011). Metro Manila flash flood of 26 September 2009: Causes, impacts and lessons learned. In A. Chayosian & K. Takeuchi (Eds.), *Large-scale floods report* (pp. 170–191). Tsukuba: Public Works Research Institute.
Noble, L. G. (1986). Politics in the Marcos era. In J. Bresnan (Ed.), *Crisis in the Philippines: The Marcos era and beyond* (pp. 70–113). Princeton, NJ: Princeton University Press.
O'Connor, J. R. (1998). *Natural causes: Essays in ecological Marxism*. New York: Guilford Press.
Office of the President of the Philippines. (1972). Official week in review. *Official Gazette of the Republic of the Philippines, 68*(32), ccxxi–ccxxiv.

Ofreneo, R. E. (1980). *Capitalism in Philippine agriculture*. Quezon City: Foundation for Nationalist Studies.

Olchondra, R. T. (2010, November 11). Manila Water to pursue plan to draw Laguna Lake water. *Philippine Daily Inquirer*. http://business.inquirer.net/money/breakingnews/view/20101111-302658/Manila-Water-to-pursue-plan-to-draw-Laguna-Lake-water

Oledan, M. T. T. (2001). Challenges and opportunities in watershed management for Laguna de Bay (Philippines). *Lakes & Reservoirs: Research & Management, 6*(3), 243–246.

Ordoñez, J. F. F., Asis, A. M. J. M., Catacutan, B. J., & Santos, M. D. (2015). First report on the occurrence of invasive black-chin tilapia *Sarotherodon melanotheron* (Ruppell, 1852) in Manila Bay and of Mayan cichlid *Cichlasoma urophthalmus* (Gunther, 1892) in the Philippines. *BioInvasions Record, 4*(2), 115–124.

Ortega, A. A. C. (2012). Desakota and beyond: Neoliberal production of suburban space in Manila's fringe. *Urban Geography, 33*(8), 1118–1143.

Ortega, A. A. C. (2016). Manila's metropolitan landscape of gentrification: Global urban development, accumulation by dispossession & neoliberal warfare against informality. *Geoforum, 70*, 35–50.

Oswin, N. (2018). Planetary urbanization: A view from outside. *Environment and Planning D: Society and Space, 36*(3), 540–546.

Pacific Consultants International. (1978). *Napindan Hydraulic Control Structure Project*. Manila: Bureau of Public Works, Ministry of Public Works, Transportation and Communications.

Palisoc, F. P. J. (1988). Fish health problems in Laguna de Bay: A preliminary report. In M. R. De los Reyes & E. H. Belen (Eds.), *First fisheries forum on fish health problems in Laguna de Bay and environs* (pp. 46–83). Los Banos, Laguna: PCARRD and Rainfed Resources Development Project.

Palma, A. L. (2015). *Knifefish infestation in Laguna de Bay*. Quezon City: BFAR-PCAARRD. http://invasivefishesbfar.blogspot.com/p/proceedings.html

Palma, A. L., Mercene, E. C., & Goss, M. R. (2005). Fish. In R. D. Lasco & M. V. O. Espaldon (Eds.), *Ecosystems and people: The Philippine millennium ecosystem assessment (MA) sub-global assessment* (pp. 115–132). College, Laguna: Environmental Forestry Programme, College of Forestry and Natural Resources, University of the Philippines Los Banos.

Palomares, M. L. D., Parducho, V. A., Bimbao, M., Ocampo, E., & Pauly, D. (2010). Philippine marine fisheries 101. *Fisheries Centre Research Reports, 22*(8), 1.

Pant, J., Barman, B. K., Murshed-E-Jahan, K., Belton, B., & Beveridge, M. (2014). Can aquaculture benefit the extreme poor? A case study of landless and socially marginalized Adivasi (ethnic) communities in Bangladesh. *Aquaculture, 418*, 1–10.

Pante, M. D. (2016). The politics of flood control and the making of Metro Manila. *Philippine Studies: Historical and Ethnographic Viewpoints, 64*(3), 555–592.

Paprocki, K. (2020). The climate change of your desires: Climate migration and imaginaries of urban and rural climate futures. *Environment and Planning D: Society and Space, 38*(2), 248–266.

Paprocki, K., & Cons, J. (2014). Life in a shrimp zone: Aqua-and other cultures of Bangladesh's coastal landscape. *Journal of Peasant Studies, 41*(6), 1109–1130.

Paraso, M. G. V., & Capitan, S. S. (2012). Vitellogenin induction and gonad abnormalities in male common carp (*Cyprinus carpio* Linnaeus) introduced to Laguna de Bay, Philippines. *Philippine Journal of Veterinary and Animal Science, 38*(1), 34–44.

Parés, M., March, H., & Saurí, D. (2013). Atlantic gardens in Mediterranean climates: Understanding the production of suburban natures in Barcelona. *International Journal of Urban and Regional Research, 37*(1), 328–347.

Pelling, M. (2003). *Natural disaster and development in a globalizing world.* London: Routledge.

Peluso, N. L., & Lund, C. (2011). New frontiers of land control: Introduction. *Journal of Peasant Studies, 38*(4), 667–681.

Peters, K. (2012). Manipulating material hydro-worlds: Rethinking human and more-than-human relationality through offshore radio piracy. *Environment and Planning A, 44*(5), 1241–1254.

Pincetl, S., Bunje, P., & Holmes, T. (2012). An expanded urban metabolism method: Toward a systems approach for assessing urban energy processes and causes. *Landscape and Urban Planning, 107*(3), 193–202.

Platon, R. R. (2001). SEAFDEC contribution to the ecological awareness of Philippine Lakes. In C. B. Santiago, M. L. Cuvin-Aralar, & Z. U. Basiao (Eds.), *Conservation and ecological management of Philippine lakes in relation to fisheries and aquaculture* (pp. 13–17). Iloilo, Los Banos, Quezon City: SEAFDEC, PCAMRD, BFAR.

Poseidon Fishing/Terry de Jesus vs. National Labor Relations Commission and Jimmy S. Estoquia. (Supreme Court 2006).

Prudham, W. S. (2005). *Knock on wood: Nature as commodity in Douglas-Fir Country.* London: Routledge.

Pullan, W. (2011). Frontier urbanism: The periphery at the centre of contested cities. *The Journal of Architecture, 16*(1), 15–35.

Rabanal, H. R., Acosta, P. A., & Delmendo, M. N. (1968). Limnological survey of Laguna de Bay—a pilot study on aquatic productivity. *Philippine Journal of Fisheries, 8*(1), 101–111.

Ranganathan, M. (2015). Storm drains as assemblages: The political ecology of flood risk in post-colonial Bangalore. *Antipode, 47*(5), 1300–1320.

Rasmussen, M. B., & Lund, C. (2018). Reconfiguring frontier spaces: The territorialization of resource control. *World Development, 101*, 388–399.

Rebotier, J. (2012). Vulnerability conditions and risk representations in Latin-America: Framing the territorializing urban risk. *Global Environmental Change, 22*(2), 391–398.

Reddy, R. N. (2018). The urban under erasure: Towards a postcolonial critique of planetary urbanization. *Environment and Planning D: Society and Space, 36*(3), 529–539.

Rey, T. J. (1987). Overview of Laguna de Bay Area Development and Management Plan. In L. C. Darvin, M. A. Mangaser, & Z. C. Gibe (Eds.), *State of development of the Laguna de Bay Area* (pp. 2–5). Los Banos, Laguna: PCARRD.

Ribot, J. C. (1998). Theorizing access: Forest profits along Senegal's charcoal commodity chain. *Development and Change*, *29*(2), 307–341.

Ribot, J. C., & Peluso, N. L. (2003). A theory of access. *Rural Sociology*, *68*(2), 153–181.

Rice, S., & Tyner, J. (2017). The rice cities of the Khmer Rouge: An urban political ecology of rural mass violence. *Transactions of the Institute of British Geographers*, *42*(4), 559–571.

Richter, H. (2001). *Seasonal variation in growth, quantitative and qualitative food consumption of milkfish, Chanos chanos (Forsskal 1775), and Nile tilapia, Oreochromis niloticus (L. 1758), in Laguna de Bay, Philippines* (Unpublished dissertation), Universitat Hohenheim.

Rickards, L., Gleeson, B., Boyle, M., & O'Callaghan, C. (2016). Urban studies after the age of the city. *Urban Studies*, *53*(8), 1523–1541.

Rivera, D. O. (2012, August 9). Manila Water posts 31% income hike. *BusinessWorld*. http://www.bworldonline.com/content.php?section=Corporate&title=Manila-Water-posts-31%-income-hike&id=56484

Rivera, F. T. (1987). Socio-cultural aspects of the fish industry around Laguna de Bay. In L. C. Darvin, M. A. Mangaser, & Z. C. Gibe (Eds.), *State of development of the Laguna de Bay area* (pp. 63–78). Los Banos: PCARRD.

Rizal, J. (2007). *El Filibusterismo*. Honolulu: University of Hawai'i Press.

RJL Martinez Fishing Corporation and/or Peninsula Fishing Corporation vs. National Labor Relations Commission et al. (Supreme Court 1984).

Robbins, P. (2001). Tracking invasive land covers in India, or why our landscapes have never been modern. *Annals of the Association of American Geographers*, *91*(4), 637–659.

Robbins, P. (2004). Comparing invasive networks: Cultural and political biographies of invasive species. *Geographical Review*, *94*(2), 139–156.

Robbins, P. (2011). *Political ecology: A critical introduction*. Hoboken, NJ: Wiley-Blackwell.

Romana-Eguia, M. R. R., & Doyle, R. W. (1992). Genotype-environment interaction in the response of three strains of Nile tilapia to poor nutrition. *Aquaculture*, *108*(1–2), 1–12.

Roy, A. (2016a). What is urban about critical urban theory? *Urban Geography*, *37*(6), 810–823.

Roy, A. (2016b). Who's afraid of postcolonial theory? *International Journal of Urban and Regional Research*, *40*(1), 200–209.

Ruaya, P. (1994, June 18). Fishing in the Bay: Are the illegal fishpen owners untouchables? *Today*, p. 9.

Salayo, N. D. (2000). *Marketing and post-harvest research (MPR) in the Philippine fisheries: A review of literature* (No. 2000-16). PIDS Discussion Paper Series.

Samonte, S. (1983, April 13). Lake law author air views. *Bulletin Today*, p. 13.
Santiago, A. E. (1990). Turbidity and seawater intrusion in Laguna Lake. *SEAFDEC Asian Aquaculture*, *12*(2), 1–4.
Santiago, A. E. (1993). Limnological behavior of Laguna de Bay: Review and evaluation of ecological status. In P. Sly (Ed.), *Laguna Lake Basin, Philippines: Problems and opportunities* (pp. 100–105). Halifax and College: Environment and Resource Management Project.
Santiago, C. B., Aldaba, M. B., Laron, M. A., & Reyes, O. S. (1988). Reproductive performance and growth of Nile tilapia (*Oreochromis niloticus*) broodstock fed diets containing *Leucaena leucocephala* leaf meal. *Aquaculture*, *70*(1–2), 53–61.
Santos-Borja, A. C. (1994). The control of saltwater intrusion into Laguna de Bay: Socioeconomic and ecological significance. *Lake and Reservoir Management*, *10*(2), 213–219.
Santos-Borja, A. C., & Nepomuceno, D. N. (2006). Laguna de Bay: Institutional development and change for lake basin management. *Lakes & Reservoirs: Research and Management*, *11*, 257–269.
Santos-Maranan, A. (1982). A dying people. In B. A. R. Foundation (Ed.), *The troubled waters of Laguna Lake* (pp. 10–18). Metro Manila: BINHI Agricultural Resource Foundation.
Saquin, H. M. (1984, May 10). LL sealane blocked by fishpens. *Bulletin Today*, p. 37.
Saraf, A. (2020). Frontiers. In *Oxford research encyclopedia of anthropology*. https://doi.org/10.1093/acrefore/9780190854584.013.145
Sarmiento, K. P., Pereda, J. M. R., Ventolero, M. F. H., & Santos, M. D. (2018). Not fish in fish balls: Fraud in some processed seafood products detected by using DNA barcoding. *Philippine Science Letters*, *11*(1), 30–36.
Schmittou, H. R., Grover, J. H., Peterson, S., Librero, A. R., Rabanal, H. R., Portugal, A. A. & Adriano, M. (1983). *Development of aquaculture in the Philippines*. Woods Hole, MA: Woods Hole Oceanographic Institution.
Schramm, S. (2016). Flooding the sanitary city: Planning discourse and the materiality of urban sanitation in Hanoi. *City*, *20*(1), 32–51.
Scott, J. C. (1998). *Seeing like a state: How certain schemes to improve the human condition have failed*. New Haven, CT: Yale University Press.
Selwyn, B. (2009). Disciplining capital: Export grape production, the state and class dynamics in northeast Brazil. *Third World Quarterly*, *30*(3), 519–534.
Selwyn, B. (2011). Beyond firm-centrism: Re-integrating labour and capitalism into global commodity chain analysis. *Journal of Economic Geography*, *12*(1), 205–226.
Seto, K. C., & Ramankutty, N. (2016). Hidden linkages between urbanization and food systems. *Science*, *352*(6288), 943–945.
Sevilla-Buitrago, A. (2015). Capitalist formations of enclosure: Space and the extinction of the commons. *Antipode*, *47*(4), 999–1020.
Sevilleja, R. C., & McCoy, E. W. (1979). *Fish marketing in Central Luzon, Philippines*. Auburn, AL: Auburn University.

Shatkin, G. (2005). Colonial capital, modernist capital, global capital: The changing political symbolism of urban space in Metro Manila, the Philippines. *Pacific Affairs*, *78*(4), 577–600.

Shatkin, G. (2008). The city and the bottom line: Urban megaprojects and the privatization of planning in Southeast Asia. *Environment and Planning A*, *40*(2), 383–401.

Shillington, L. J. (2013). Right to food, right to the city: Household urban agriculture, and socionatural metabolism in Managua, Nicaragua. *Geoforum*, *44*, 103–111.

Silver, J. (2015). Disrupted infrastructures: An urban political ecology of interrupted electricity in Accra. *International Journal of Urban and Regional Research*, *39*(5), 984–1003.

Simon, D. (2008). Urban environments: Issues on the peri-urban fringe. *Annual Review of Environment and Resources*, *33*, 167–185.

Simon, G. L. (2014). Vulnerability-in-production: A spatial history of nature, affluence, and fire in Oakland, California. *Annals of the Association of American Geographers*, *104*(6), 1199–1221.

Simone, A. (2004). People as infrastructure: Intersecting fragments in Johannesburg. *Public Culture*, *16*(3), 407–429.

Simone, A. (2010). *City life from Jakarta to Dakar: Movements at the crossroads*. London: Routledge.

Simone, A. (2019). *Improvised lives: Rhythms of endurance in an urban south*. Cambridge, UK: Polity.

Simone, A. (2020). To extend: Temporariness in a world of itineraries. *Urban Studies*, *57*(6), 1127–1142.

Sly, P. (Ed.). (1993). *Laguna Lake Basin, Philippines: Problems and opportunities*. Halifax and College: Environmental and Resource Management Project.

Smith, N. (2008). *Uneven development: Nature, capital, and the production of space*. Athens: University of Georgia Press.

Sneddon, C. (2007). Nature's materiality and the circuitous paths of accumulation: Dispossession of freshwater fisheries in Cambodia. *Antipode*, *39*(1), 167–193.

Sneddon, C. (2015). *Concrete revolution: Large dams, Cold War geopolitics, and the US Bureau of Reclamation*. Chicago: University of Chicago Press.

Sneddon, C., & Fox, C. (2012). Inland capture fisheries and large river systems: A political economy of Mekong fisheries. *Journal of Agrarian Change*, *12*(2–3), 279–299.

SOGREAH. (1991). *Environmental assessment of Laguna de Bay: Final report*: Paris: France Ministere Des Affaires Estrangeres Direction Du Development et de la Cooperation Scientifique, Technique et Educative.

Spoehr, A. (1984). Change in Philippine capture fisheries: An historical overview. *Philippine Quarterly of Culture and Society*, *12*(1), 25–56.

Star, S. L. (1999). The ethnography of infrastructure. *American Behavioral Scientist*, *43*(3), 377–391.

Starosta, G. (2010). Global commodity chains and the Marxian law of value. *Antipode, 42*(2), 433–465.

Stevenson, J. R., & Irz, X. (2009). Is aquaculture development an effective tool for poverty alleviation? A review of theory and evidence. *Cahiers Agricultures, 18*(2–3), 292–299.

Stonich, S. C., Bort, J. R., & Ovares, L. L. (1997). Globalization of shrimp mariculture: The impact on social justice and environmental quality in Central America. *Society & Natural Resources, 10*(2), 161–179.

Strauss, K. (2013). Unfree again: Social reproduction, flexible labour markets and the resurgence of gang labour in the UK. *Antipode, 45*(1), 180–197.

Surtida, A. P. (2000). Middlemen: The most maligned players in the fish distribution channel. *SEAFDEC Asian Aquaculture, 22*(5), 21–22.

Swyngedouw, E. (1996). The city as a hybrid: On nature, society and cyborg urbanization. *Capitalism Nature Socialism, 7*(2), 65–80.

Swyngedouw, E. (2004). *Social power and the urbanization of water: Flows of power.* Oxford: Oxford University Press.

Swyngedouw, E. (2006). Circulations and metabolisms: (Hybrid) natures and (cyborg) cities. *Science as Culture, 15*(2), 105–121.

Swyngedouw, E., & Heynen, N. C. (2003). Urban political ecology, justice and the politics of scale. *Antipode, 35*(5), 898–918.

Tabbu, M., Lijauco, M., Eguia, R., & Espegadera, C. (1986). Polyculture of bighead carp, common carp and Nile tilapia in cages in Laguna Lake. *Fisheries Research Journal of the Philippines, 11*(1–2), 13–20.

Tabios, G. Q. (2019). Urban dimensions of floodings and holistic flood risk management: Case of Pasig-Marikina River Basin in Metro Manila. In T. S. E. Tadem (Ed.), *A study of the implications of federalism in the National Capital Region and considerations for forming the Federal Administrative Region* (pp. 146–166). Quezon City: UP CIDS.

Tabios, G., & David, C. (2004). Competing uses of water: Cases of Angat reservoir, Laguna Lake and groundwater systems of Batangas City and Cebu City. In A. C. Rola, H. A. Francisco, & J. P. T. Liguton (Eds.), *Winning the water war: Watersheds, water policies and water institutions* (pp. 105–131). Makati City: PIDS and PCARRD.

Tadem, T. S. E. (2013). Philippine technocracy and the politics of economic decision making during the martial law period (1972–1986). *Social Science Diliman, 9*(2), 1–25.

Takahashi, A., & Fegan, B. (1983). Two views of the kasama-lessee shift in Bulacan: An exchange. In A. J. Ledesma, P. Q. Makil, & V. A. Miralao (Eds.), *Second view from the paddy*. Quezon City: Institute of Philippine Culture, Ateneo de Manila University.

Tamayo-Zafaralla, M., Santos, R. A. V., Orozco, R. P., & Elegado, G. C. P. (2002). The ecological status of lake Laguna de Bay, Philippines. *Aquatic Ecosystem Health & Management, 5*(2), 127–138.

Tan, R. L., Alvaran, T. A. C., Villamor, B. B., & Tan, I. M. A. (2010). Cost and return analysis of fishpen operation in Laguna de Bay and the economic implication of "Zero Fishpen Policy". *Journal of Environmental Science and Management*, 13(2) 14–26.

Tapia, M., & Zambrano, L. (2003). From aquaculture goals to real social and ecological impacts: Carp introduction in rural central Mexico. *AMBIO: A Journal of the Human Environment*, 32(4), 252–258.

Taylor, M. (2007). Rethinking the global production of uneven development. *Globalizations*, 4(4), 529–542.

Tiambeng, R. G. (1992). *A study on fish marketing system in Navotas Fishing Port and Fish Market, Navotas, Metro Manila* (Unpublished thesis), Colegio de San Juan de Letran, Manila.

Tinker, I. (1997). *Street foods: Urban food and employment in developing countries*. Oxford: Oxford University Press.

Tolentino, M. C. M. (2010). Child labor in Navotas, Philippines: Social construction of reality. In O. F. von Feigenblatt (Ed.), *Development and conflict in the 21st Century* (pp. 128–139). Bangkok: JAPSS Press.

Torres, E. B., Pabuayon, I. M., & Salayo, N. D. (1987). *Market structure analysis of fish distribution channels supplying Metro Manila*. Joint project by the Asian Fisheries Social Science Research Network and the Department of Agricultural Economics, University of the Philippines Los Banos, Laguna.

Toufique, K. A., & Gregory, R. (2008). Common waters and private lands: Distributional impacts of floodplain aquaculture in Bangladesh. *Food Policy*, 33(6), 587–594.

Tribdino, R. B. (1995, September 21). As over 100 pens seen to be affected by dismantling. *BusinessWorld*, p. 11.

Tribdino, R. B. (1996, January 19). Laguna Lake fishermen want first crack in issuance of fishing permit (with inevitable competition from big business). *BusinessWorld*, p. 6.

Tsing, A. L. (2003, November 29). Natural resources and capitalist frontiers. *Economic and Political Weekly*, 5100–5106.

Tsing, A. L. (2005). *Friction: An ethnography of global connection*. Princeton, NJ: Princeton University Press.

Tsing, A. L. (2015). *The mushroom at the end of the world: On the possibility of life in capitalist ruins*. Princeton, NJ: Princeton University Press.

Turgo, N. (2012). "Bugabug ang dagat" (rough seas): Experiencing Foucault's heterotopia in fish trading houses. *Social Science Diliman*, 8(1), 31–62.

Turner, F. J. (1920). *The frontier in American history*. New York: Holt. (Original work published 1893)

United Nations. (1970). *Feasibility survey for the hydraulic control of the Laguna de Bay complex and related development activities: Investment opportunity analyses of Laguna de Bay hydraulic control*. Vancouver: T. Ingledow and Associates.

Vandergeest, P., Flaherty, M., & Miller, P. (1999). A political ecology of shrimp aquaculture in Thailand 1. *Rural Sociology*, 64(4), 573–596.

Vandergeest, P., & Peluso, N. L. (1995). Territorialization and state power in Thailand. *Theory and Society*, *24*(3), 385–426.
VDA Fish Broker and/or Venerando Alonzo vs. National Labor Relations Commission, Ruperto Bula and Virgilio Salac. (Supreme Court 1993).
Veuthey, S., & Gerber, J. F. (2012). Accumulation by dispossession in coastal Ecuador: Shrimp farming, local resistance and the gender structure of mobilizations. *Global Environmental Change*, *22*(3), 611–622.
Villadolid, D. (1933). Some causes of depletion of certain fishery resources of Laguna de Bay. *Natural and Applied Science Bulletin*, *3*, 251–255.
Villadolid, D. (1934). Kanduli fisheries of Laguna de Bay Philippine Islands: Remedial and regulatory measures for their rehabilitation. *Philippine Journal of Science*, *54*(4), 545–552.
Villaluz, D. K. (1950). *Fish farming in the Philippines*. Manila: Bookman.
Walker, P., & Fortmann, L. (2003). Whose landscape? A political ecology of the "exurban" Sierra. *Cultural Geographies*, *10*(4), 469–491.
Walker, G., Whittle, R., Medd, W., & Walker, M. (2011). Assembling the flood: Producing spaces of bad water in the city of Hull. *Environment and Planning A*, *43*(10), 2304–2320.
Warren, J. F. (2013). A tale of two decades: Typhoons and floods, Manila and the provinces, and the Marcos years. *The Asia-Pacific Journal: Japan Focus*, *11*(43/3).
Warren-Rhodes, K., & Koenig, A. (2001). Escalating trends in the urban metabolism of Hong Kong: 1971–1997. *AMBIO: A Journal of the Human Environment*, *30*(7), 429–438.
Watts, M. (2005). Commodities. In P. Cloke, P. Crang, & M. Goodwin (Eds.), *Introducing human geographies* (pp. 527–546). Oxon, OX: Hodder Arnold.
Watts, M. J. (2018). Frontiers: Authority, precarity, and insurgency at the edge of the state. *World Development*, *101*(C), 477–488.
Wegerif, M. C., & Wiskerke, J. S. (2017). Exploring the staple foodscape of Dar es Salaam. *Sustainability*, *9*(6), 1081.
Weis, T. (2010). The accelerating biophysical contradictions of industrial capitalist agriculture. *Journal of Agrarian Change*, *10*(3), 315–341.
Williams, R. (1973). *The country and the city*. New York: Oxford University Press.
Yamane, A. (2009). Climate change and hazardscape of Sri Lanka. *Environment and Planning A*, *41*(10), 2396–2416.
Yap, W. G. (1999). *Rural aquaculture in the Philippines*. Bangkok: Food and Agriculture Organization.
Yates, J. S., & Gutberlet, J. (2011). Reclaiming and recirculating urban natures: Integrated organic waste management in Diadema, Brazil. *Environment and Planning A*, *43*(9), 2109–2124.
Yosef, S. (2009). *Rich food for poor people: Genetically improved tilapia in the Philippines*. Washington, DC: International Food Policy Research Institute.
Zafaralla, M. T., Barril, C. R., Santos-Borja, A. C., Manalili, E. V., Dizon, J. T., Sta. Ana, J. G., & Aguilar, N. O. (2005). Water resources. In R. D. Lasco & M. V. O. Espaldon (Eds.), *Ecosystems and people: The Philippine millennium*

ecosystem assessment (MA) sub-global assessment (pp. 63–114). College, Laguna: Environmental Forestry Programme, College of Forestry and Natural Resources, University of the Philippines Los Banos.

Zeiderman, A. (2012). On shaky ground: The making of risk in Bogotá. *Environment and Planning A*, *44*(7), 1570–1588.

Zhang, Y. (2013). Urban metabolism: A review of research methodologies. *Environmental Pollution*, *178*, 463–473.

Zimmer, A. (2010). Urban political ecology: Theoretical concepts, challenges, and suggested future directions. *Erdkunde*, *64*(4) 343–354.

Zimmer, A. (2015). Urban political ecology "beyond the West": Engaging with South Asian urban studies. In R. L. Bryant (Ed.), *The international handbook of political ecology* (pp. 591–603). Cheltenham: Edward Elgar Publishing.

INDEX

1952 Plan for the Drainage of Manila and Suburbs, 145–46
1972 Manila and Suburbs Flood Control Plan, 146

access (to resources), 2, 18–19, 21, 26, 56
access of fisherfolk to Laguna Lake, 56, 63–71, 106
agrarian political economy, 20
algae blooms, 6, 47, 62, 90–91, 128, 137
Angat Dam, 49
aquaculture: agrarian change, 71–73 (see also Kalinawan; Navotas); capitalist aquaculture, 59, 86, 95, 159; conflicts, 49–50; expansion, 42, 54; fixity of, 89, 154; hazards (see flood; typhoon); interaction with capture fisheries, 6, 8; introduction, 37–38; limitations, 125; production, 2–8; reliance on nature, 90; social relations transformation, 77–78; as solution to more efficient fish production, 25; as source of cheap fish, 130; technologies, 40. See also fishcage; fish corral; fishpen; fishpond
Aquino, Benigno, III, 158
ayungin, 40

backflow, 6, 45–48, 62, 68, 87–90. See also saltwater intrusion
batilyo, 118–21
bighead carp, 8, 89, 107, 110, 124–28, 132–35. See also Imelda; mamali; maya-mayang tabang

biya, 40
brokers, 111–20
bulungan, 112, 116
burning of Laguna Lake in July 1982, 53

Calabarzon Project, 49
Calabarzon region, 8–9
Calamba (city), 23
capitalism, 16–18; capitalist enclosure, 72; capitalist urbanization, 18; overcoming nature, 20. See also commodity frontier
capture fisheries: coexistence with aquaculture, 71–73 (see also Kalinawan; Navotas); expansion of, 56; gears used in, 38–39 (see also cast net; drag seine; drift long line; gill net; motorized push net); interaction with aquaculture, 6, 8, 37–38, 40, 42–55; mobility of, 65
cast net (dala), 39
city, 13–19, 72, 74, 157; metabolic relations with frontier, 22, 24, 125. See also frontier making; Metro Manila; urbanization
clientelism, 116–17, 122. See also suki
clown featherback. See knifefish
commodity chain, 106. See also value chain
commodity deepening, 55
commodity frontier, 17, 55
commodity widening, 55
construction of flood control infrastructure, 146
contractualization, 121
cooperatives, 70

demolition, 53–54, 67
drag seine (pukot), 38–39, 63, 73, 111, 115
drift long line (kitang), 39, 99
Duterte, Rodrigo, 160–61

ecologies, 10–12, 14, 18; agrarian ecologies, 54; political ecologies, 82, 162; urban ecologies, 20. *See also* city; flow; frontier
edge: as a location, 10, 44, 97, 158–59, 161; as a relation, 10–11; uncertainty of, 10. *See also* urban edge
enclosure, 56–57
environmental management, 3
eutrophic property of Laguna Lake, 6, 44–45, 62, 89–90, 128
evictions, 152–53

fish ball, 135
Fish Brokers Association, 121
fishcage: cage nursery, 74, 76–77, 115; cage producer, 154–55; cage production, 77, 83, 94–95; coexistence with fishpen, 37, 41
fish consumption, 129–30
fish corral, 41
fisherfolk, 39, 77
Fishery Zoning and Management Plan (ZOMAP), 67–69, 84–85
fishpen, 37, 123; coexistence with fish cage, 37, 41; controversy, 62–65; demolition of, 53–54; development, 42–43; expansion, 58; fishpen caretaker, 92–93; fishpen laborer, 92–94
fishpen rush, 59, 62
fishpond, 59, 61
fish production, 45, 56, 68–69, 73, 107, 126
Floating Cage Project, 41
flood, 2–5, 81
flood control, 34–36, 140–43, 146–50
flow, 12–13, 105–8; food flow, 125, 130; fish flow, 136–39; water flow, 140–49
fluid urbanism, 5
food, 12–13, 125
foreman, 118–21
frontier, 2–4, 14–19, 32–34, 49–51
frontier landscapes, 3
frontier making, 10, 15–17, 19, 21, 33–34, 56, 80, 123, 159
frontier urbanism, 3, 4, 19, 22, 124, 142

galunggong, 130, 136
gill net (pante), 39, 83, 114

habagat, 1, 169n1

Imelda, 125–26, 134
infrastructure, 19, 21–22, 32, 45; paradoxical character of, 142–43, 148; as metabolic flow control, 48, 51, 146–47; as solution to metabolic problems of city, 151–53, 161–62. *See also* Mangahan Floodway; Napindan Hydraulic Control Structure
invasives, 96. *See also* janitor fish; knifefish

janitor fish, 96–97

kalaso, 135
Kalinawan (barangay in Binangonan municipality), 73
kanduli, 39
kasama system, 73
knifefish, 1–2, 96–101, 135

Lagumbay, Wenceslao, 32
Laguna de Bay Fishpen Development Program, 70–71
Laguna de Bay Master Plan, 85
Laguna Lake, 1–10, 31–34, 163n3; as dead or dying, 160; decline in productivity of, 62; development, 34–36; experimental farm in, 40–41; in fiction, 31–32; fishing regulations in, 63; infrastructure intervention in, 45–47; limnological studies on, 45–47; overexploitation and underutilization of, 32; poor water condition in, 89; as stormwater sink, 148–51; unrest and violence in, 64–66; as urban resource frontier, 33; value chain of, 111; as water resource, 35, 49
Laguna Lake Cooperative Development Program, 70
Laguna Lake Development Authority (LLDA), 6, 34–35, 40–42, 70, 85–86
Laguna Lakeshore Expressway Dike (LLED), 158–59
lasang gilik, 91
Lavides, Vicente, Jr., 82

lizard fish, 135. *See also* kalaso
LLDA, 6, 34–35, 40–42, 70, 85–86
LLED, 158–59
Looc Fish Pen Demonstration Project, 40–41

Macapagal-Arroyo, Gloria, 51, 130, 152
Malabon (city of), 41, 59, 108, 111–12, 115
mamali, 125–26
Mangahan Floodway, 148
Manila (City of), 5, 145, 163n2
Manila Bay, 5–6, 59, 128, 159
Manila catfish, 39. *See also* kanduli
Marcos, Ferdinand, 35–36, 64, 67, 70, 146–47, 153
Marcos, Imelda, 35, 126, 146–47
Marikina River, 96, 140, 146–48, 150
materiality of nature, 19–21, 26, 62, 80–81, 86, 89, 94, 102, 132
maya-mayang tabang, 125–26, 132
Metro Manila, 1, 5–9, 163n1; development of flood control infrastructure in 147–52; fish consumption in, 130, 134–35; flooding in, 140, 143; population growth, 36; socializing chains with Laguna Lake, 111–12, 115; water development in, 49–50
militarization, 57, 63–64
milkfish, 2, 8, 50–51, 89, 107, 110
motorized push net (sakag, suro), 38–39, 73, 76–77
Muntinlupa, 23

Napindan Hydraulic Control Structure (NHCS), 47–48, 155–56
Navotas (barangay in Cardona municipality), 76
Navotas (City of), 41, 59, 61, 59, 108, 111–12, 115, 117
Navotas Fish Port Complex, 23, 105–8, 117
NHCS, 47–48, 155–56

Pasig River, 5–6, 45, 47, 87, 140, 145–48, 156
peri-urban, 10–11, 17. *See also* edge
Philippine Fisheries Development Authority, 108
poaching, 66, 93
pollution, 50

Polyculture Development Program, 43–44
practices and experiences, 19, 22
Putatan water plant, 50

Ramos, Fidel, 85
Ramos, Victor, 85
red snapper, 132. *See also* maya-mayang tabang
reproduction squeeze, 72
resource frontier, 3–4, 15–17, 32–33, 55–56, 80–81, 101, 151–59
retailer, 116, 131, 133–34
Rizal province, 23
round scad, 130, 136. *See also* galunggong

sailfin catfish, 96–97. *See also* janitor fish
saltwater intrusion, 6, 45–48, 62, 68, 87–90
Samahang Mandaragat ng Lawa ng Laguna, Ink., 64
Samahan ng Nagkakaisang Batilyo-NFL, 121
scavengers, 117–18
SEAFDEC, 43, 44, 48, 73
sewer and water system development for Metro Manila, 143–45
silver perch, 40. *See also* ayungin
Société Grenobloise d'Etudes et d'Application Hydrauliques, 47
socionatural hybridity, 20
Southeast Asian Fisheries Development Center (SEAFDEC), 43–44, 48, 73
southwest monsoon, 1, 169n1. *See also* habagat
suki, 116–17, 122

Taguig, 23
tilapia, 8, 51, 107, 110, 130; Genetically Improved Farmed Tilapia, 44; Mozambique tilapia, 43; Nile tilapia, 43
trade union, 121
trawl line, 99. *See also* drift long line (kitang)
typhoon: Basyang (Conson), 84; Kading (Rita), 81; Katring (Teresa), 85; Mameng (Sibyl), 85; Ondoy (Ketsana), 1, 84; Rosing (Angela), 85; Weling (Lola), 81; Yaning (Ora), 81

unmapping, 16
urban commodity flow, 3

urban connections, 4
urban development, 3
urban edge, 9–11
urban frontier, 4, 18–19
urbanization, 3, 13–19, 21–22, 33–36, 158, 161–62
urban metabolism, 3–4, 11–13, 18, 21–22, 105–6, 138–39, 161
urban political ecology, 11–12, 17, 20, 155, 162
urban provisioning, 2, 106, 108, 123–24, 131, 142, 153, 158
urban socioecology, 2, 10
urban sustainability, 11

value chain, 108, 111–12, 115, 117, 122, 153, 161–62

water, 12; commodification of, 87–89; development of, 49; development of water management system, 144, 150–51; effects of fluctuating water condition, 94–95, 128; good vs bad water, 13; materiality of, 20; new projects for Manila Manila water supply, 161; privatization of, 50; urbanization of, 18
white goby, 39. *See also* biya
wholesalers, 111, 116–17, 120

yellowfin tuna, 134

ZOMAP, 67–69, 84–85

Founded in 1893,
UNIVERSITY OF CALIFORNIA PRESS
publishes bold, progressive books and journals
on topics in the arts, humanities, social sciences,
and natural sciences—with a focus on social
justice issues—that inspire thought and action
among readers worldwide.

The UC PRESS FOUNDATION
raises funds to uphold the press's vital role
as an independent, nonprofit publisher, and
receives philanthropic support from a wide
range of individuals and institutions—and from
committed readers like you. To learn more, visit
ucpress.edu/supportus.

Lightning Source UK Ltd.
Milton Keynes UK
UKHW010645151022
410492UK00006B/473